TABLE OF CONTENTS

Page

ACRONYMS

AT&L	Acquisition, Technology and Logistics
AVGAS	Aviation Gas
BAMS	Broad Area Maritime Surveillance
C4ISR	Command, Control, Communication, Computer, Intelligence, Surveillance and Reconnaissance
CARL	Combined Armed Research Library
CBO	Congressional Budget Office
COCOM	Combatant Command
DAU	Defense Acquisition University
DARO	Defense Airborne Reconnaissance Organization
DEPSECDEF	Deputy Secretary of Defense
DHS	Department of Homeland Security
DOD	Department of Defense
DSB	Defense Science Board
EO	Electro-Optical
ER/MP	Extended Range/Multipurpose
FCS	Future Combat System
FYDP	Future Years Defense Plan
GAO	Government Accountability Office
HAE	High Altitude Endurance
HASC	House Armed Services Committee
HMMWV	High Mobility Multi-purpose Wheeled Vehicles
IOC	Initial Operating Capability

IR	Infrared
ISR	Intelligence, Surveillance and Reconnaissance
JCIDS	Joint Capabilities Integration and Development System
JP	Jet Propulsion
JROC	Joint Requirements Oversight Council
JUAS CoE	Joint Unmanned Aircraft System Center of Excellence
MILCON	Military Construction
MRB	Material Review Board
MTI	Moving Target Indicator
NASA	National Aeronautics and Space Administration
NATO	North Atlantic Treaty Organization
NDAA	National Defense Authorization Act
NOAA	National Oceanic and Atmospheric Administration
O&M	Operations and Maintenance
OEF	Operation Enduring Freedom
OIF	Operation Iraqi Freedom
OSD	Office of the Secretary of Defense
PPBE	Planning, Programming, Budgeting and Execution
RDT&E	Research, Development, Test and Evaluation
SAR	Synthetic Aperture Radar
SECDEF	Secretary of Defense
TF	Task Force
UAS	Unmanned Aircraft System
UAV	Unmanned Aerial Vehicle
US	United States

USA	United States Army
USAF	United States Air Force
USN	United States Navy

ILLUSTRATIONS

TABLES

CHAPTER 1

INTRODUCTION

In the Department of Defense's (DOD) fiscally constrained environment, the fight

for resources is usually a zero sum game. What is gained for one program comes at the

cost of another program unless the DOD budget expands. In the near future, funding

related to Operation Iraqi Freedom (OIF) and Operation Enduring Freedom (OEF) will

come to an end, and the DOD will have tough fiscal decisions to make. On 30 January

2009, the new administration under President Barack H. Obama asked the military's Joint

Chiefs of Staff to cut the DOD budget submission for Fiscal Year 2010 (FY10) by more

than ten percent.[1] With new national priorities aimed at saving money and reducing

waste, all government programs can expect a thorough evaluation for potential budget

cuts. The acquisition of medium and high unmanned aircraft systems (UAS) is not

exempt from this scrutiny. Based upon increasing pressure for fiscal constraint, two

questions need to be addressed. First, are there economic efficiencies that could be

achieved by appointing a lead organization to direct the procurement of these unmanned

systems? Second, are the requirements between current medium and high altitude UAS

similar enough to combine some of these airframes into one platform?

The procurement of UAS, previously referred to as unmanned aerial vehicles

(UAV), has dramatically increased since the onset of OEF and OIF. In the medium to

high altitude UAS category alone, DOD spends over two billion dollars every year to

research, develop, procure and operate these highly sophisticated aircraft and their related

equipment and systems.[2] Of note, the capabilities of many of these medium to high

altitude UASs are very similar. In an effort to rapidly field UAS capabilities to support

1

current military operations, there appears to be numerous cost inefficiencies incurred by all military services. Is the DOD failing to recognize potential savings from current UAS research and procurement practices that could be redirected to other acquisition programs? Perhaps these funds could be reinvested to acquire even more UAS platforms in high demand?

For over 30 years, medium to high altitude UAS programs have played an ever increasing role in military operations. In 1979, the U.S. began its first major UAV program, the Aquila. This small propeller driven aircraft was designed to provide ground forces with real-time information to assist artillery, designate targets and survive against air defenses. Cost overruns and technical complications ended this program in 1987.[3] The ability to create an unmanned weapon system program proved too difficult with the technology available at that time and the funding limitations.

In 1986, the U.S. Navy cooperated with Israel to procure nine Pioneer UAV systems. Like the Army Aquila, this was a small propeller driven aircraft designed to support naval gunfire and Marine Corps forces on the ground. Numerous setbacks, including ship-borne recovery and electro-magnetic interference caused several fielding problems for the Pioneer, yet the Navy persevered through these issues and the Pioneer developed into a mature system that saw combat in Desert Storm, Somalia, and Bosnia.[4]

In 1993, the U.S. Air Force entered the medium altitude UAV arena in a joint venture with the U.S. Navy designated the Medium Range UAV. This aircraft was envisioned to precede strike aircraft behind enemy lines and loiter to provide battle damage assessments. The Air Force built the sensor payloads and communications equipment, while the Navy constructed the airframe and propulsion. Both services experienced considerable setbacks. In the end, the Air Force's prototype payload did not

fit into the Navy's designed airframe and the program was cancelled.[5] A dual

procurement approach led to this unfortunate outcome. No service was the overall

project manager, leading both services to develop their respective Medium Range UAV

components independently. Perhaps if one organization was in charge, the physical

airframe oversight would have been rectified through program oversight before the merge

of payload and airframe.

From these past efforts, DOD recognized the incredible potential that medium and

high altitude UAS held for all uniformed services. A pilotless aircraft presented several

benefits to include additional space for equipment, removal of human body limits on

performance and the avoidance of loss of life. As intelligence, surveillance and

reconnaissance (ISR) platforms, their endurance and evolving sensors made them optimal

military reconnaissance vehicles. Arming these remotely piloted airframes transformed

them into strike aircraft able to prosecute a variety of targets from terrorists to enemy

tank columns.

Currently, there are five medium and high altitude UAS in the production or

research and development phase that are planned and programmed to play significant

roles for all three military services. These UAS are the MQ-1 Predator, MQ-9 Reaper

and RQ-4 Global Hawk produced by the Air Force, along with the MQ-1C Sky Warrior

and the Broad Area Maritime Surveillance (BAMS) produced by the Army and Navy

respectively. However, redundant capabilities and lack of authority among these systems

and their respective owners highlighted the need for a thorough and objective look at

DOD acquisition strategy and the appointment of a single acquisition organization for medium and high altitude UAS.

The Office of the Secretary of Defense for Acquisition, Technology and Logistics (OSD AT&L) conducted and released such an assessment on two similar systems in 2008. OSD AT&L estimated that consolidation of the Air Force MQ-1B Predator and Army MQ-1C Sky Warrior programs alone, with a common depot could save $400 - $600 million over the life of both programs.[6] This does not include benefits realized from common basing, training and sustainment which feel outside the realm of this study.

Likewise, the Government Accountability Office (GAO) recognized similar UAS improvement areas since 2004. The GAO cited the DOD's lack of strategic UAS planning and UAS authority was detrimental to the management of UAS programs.[7] Without correcting both UAS improvement areas, redundancy and future force structure complications could be expected.

Yet there was still no single directive authority over UAS development and procurement with the power to mandate this change recognized by the GAO and DOD. As these systems mature, the ability to maximize financial savings and operational capabilities will decline. By establishing a good business case from the onset, inefficiencies are reduced. Early programmatic intervention increases the probability of delivering the required warfighter capabilities within the programmed funding.[8] With a large quantity of these medium and high altitude UAS entering operational status over the next several years, DOD cannot afford to overlook the economic benefits of appointing one organization to direct the acquisition of all medium to high altitude UAS.

Primary and Secondary Research Questions

The primary research question of this thesis addresses the organizational and economic facets of the UAS issue. Are there economic efficiencies that could be achieved by appointing a lead service or agency to direct the procurement of the five medium to high altitude UAS systems under review: MQ1-B Predator (USAF), MQ-1C Sky Warrior (USA), MQ-9 Reaper (USAF), RQ-4 Global Hawk (USAF), and the Broad Area Maritime Surveillance-BAMS (USN)?

Two secondary research questions address the current UAS requirements influencing the direction of acquisition efforts. First, are there similar mission requirements that support the potential to combine some of the UAS systems into a single system, administered by one agency, service or organization? Second, what are these differences in UAS requirements for each service driving divergent acquisition programs?

Significance

The appointment of a single acquisition authority could potentially save hundreds of millions to billions of dollars that DOD could re-invest for the procurement of even more medium to high altitude UAS. Another beneficial option would be to allocate these cost savings to other high priority acquisition programs or requirements within the Defense Department. A single acquisition authority could also ensure commonality and reduce redundant capabilities on all current and future UAS procurement.

A single acquisition authority is not an untried approach for the DOD in regard to the procurement of UAS. In 1988, the Pentagon introduced the *DOD Joint Unmanned Aerial Vehicle (UAV) Master Plan-1988*.[9] This master plan outlined the basis of a congressionally mandated consolidation of all remotely piloted vehicle programs from the services into one joint organization. In 1989, this organization became known as the UAV Joint Project Office at the direction of DOD Director of Defense Research and Engineering.[10] The Navy was appointed the executive agent of this joint organization to exert authority on all UAV programs and eliminate duplication of effort.[11] But the duplication of effort was never solved. Four years later, the GAO determined that the DOD UAV Master Plan of 1988 contained four main flaws

> (1) did not eliminate duplication, (2) continued to permit the proliferation of single-service programs, (3) did not adequately consider cost savings potential from manned and unmanned aircraft trade-offs, and (4) did not adequately emphasize the importance of common payloads among different UAV platforms.[12]

In 1993, the DOD UAV Joint Project Office and naval executive agency was replaced by the Defense Airborne Reconnaissance Office (DARO) to assume the management, oversight and coordination of all DOD reconnaissance programs, manned or unmanned.[13] DARO also owned all DOD UAS funding and stayed in existence for approximately five years before a poor management approach and lack of results in fielding operational UAS brought an end to its existence. Ultimately, nine years of a single DOD acquisition owner of UAS funds accomplished little to streamline the procurement of UAS systems in the DOD inventory.

Following the aforementioned failures, DOD dissolved the DARO organization in 1998 and UAS program ownership and acquisition authority were returned to the respective services. The Assistant Secretary of Defense for Command, Control, Communications and Intelligence assumed the oversight responsibility on behalf of the

Secretary of Defense.[14] Yet, the services maintained funding authority for all UAS programs in their programmed budgets.

In October 2001, the DOD established a Joint UAS Planning Task Force within the Office of the Undersecretary of Defense for Acquisition, Technology and Logistics (OSD AT&L) in order to provide direction and guidance for future UAS investments. This organization subsequently evolved into the current DOD UAS Task Force in October of 2007.[15] The two changes since 2001 reassigned oversight of UAS acquisition to task forces within OSD AT&L, but none of the new changes addressed who had primary fiscal authority. It appeared that since a single UAS acquisition authority did not work under naval executive agency and DARO, subsequent DOD restructuring efforts avoided changing funding ownership. Yet, there is a significant difference between 2009 and 1993.

A look at the current contributions of UAS to the operational environment showcased an exponential climb in flying hours when compared to the mid-90s. The graphic below reflects only the medium and high altitude UAS programs and the combination of flying hours for all services. It is important to note that from 2001 and especially from 2003, most of the flying hours are combat related. The exponential climb in operational use identified a trend that would only continue to rise based on the programmed procurement plan for the five aforementioned UAS. The number of medium and high altitude UAS and their combat contributions in 2008 highlights an exponential trend compared to the mid-1990s.

DOD was not the only branch of government interested in the ramifications of the sudden rise in UAS involvement. Congress fully understood the pending boom in UAS contributions to future military operations. They ensured through directive language that the DOD took efforts to effectively manage and transition to the rapidly expanding arena of unmanned systems: that ―By 2010, one third of the aircraft in the operational deep strike force should be unmanned.‖[16] The 2006 John Warner National Defense Authorization Act (NDAA) amended the 2001 mandate by removing the one third restrictions from weapon system programs already under production or with funds already appropriated. It did mandate consideration of a UAS option for all new aircraft systems entering the DOD force structure.[17]

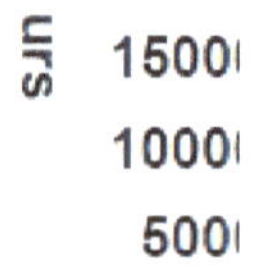

hrs
1500ı
1000ı
500ı

Figure 1.   Medium and High Altitude UAS Flying Hours 1996 – 2008
Source: DOD UAS Task Force Briefing 9 January 2009, Unmanned Aircraft Systems, (Washington, DC: Government Printing Office, 2009), 5.

Funding was not the only area of concern with regards to the exponential growth in medium and high altitude UAS. In July of 2005, the Joint Requirements Oversight Council (JROC) created the Joint UAS Center of Excellence (JUAS CoE) and Joint UAS

Material Review Board (MRB) to address UAS capability issues and to prioritize solutions within the DOD. The charter of these two organizations was envisioned to address interoperability and duplication of effort issues facing the DOD in the procurement and development of UAS programs. Even though these two relatively new organizations had greater visibility into the entire DOD UAS programmatic structure and access to very key DOD and JCS decision-makers, there was still one key weakness not addressed. The DOD created these two entities to guide the respective services in UAS acquisition, planning, prioritization and execution.[18] Yet, neither of these Joint organizations possessed the authority to mandate any of their suggestions. Nor do either of them own funding profiles associated with service UAS programs. The JUAS CoE and MRB could only advise and recommend to the military services and JROC recommended courses of action.[19] This authority shortfall allowed each respective service to continually proceed down separate procurement lanes in their efforts to meet individual service requirements. This lack of authority would become very apparent in the next few years.

On 22 April 2008, the US Navy announced the selection of Northrop Grumman as the contract award winner for the Broad Area Maritime System (BAMS), a high altitude UAS system designed for maritime reconnaissance worth an estimated $1.2 billion dollars according to the 2008 President's Budget Request.[20] The most notable trait of the Northrop Grumman contract is that the BAMS platform utilizes the same airframe as the Air Force RQ-4 Global Hawk currently fielded in combat and in full rate production.[21] The significance of the BAMS platform is it appeared to be a perfect candidate for a single organization acquisition authority as the Navy could integrate with the ongoing Air

Force production line.[22] However, even with service specific payloads, use of a common contract representing the requirements of multiple services should have been utilized to capitalize on economies of scale.

For example in 1983, AM General won the contract to produce High Mobility Multi-purpose Wheeled Vehicles (HMMWV) for the US Army.[23] These vehicles utilized the same basic chassis, but contained 15 different configurations from weapon platforms to utility vehicles. This contract for nearly 55,000 vehicles was administered under one contract, representing the requirements for the Army, Navy, Air Force and Marine Corps.[24] Procuring more vehicles with one contract can lower per unit costs as fixed costs can be spread across more units. In addition, vendors can benefit from larger contracts as their suppliers usually drop freight, shipping and wholesale cost for bigger purchases, thereby passing some savings on to the unit cost.[25] Unfortunately, this did not occur for the Global Hawk and BAMS; a single acquisition authority for UAS could have mandated or restructured the contract with Northrop Grumman to gain economic savings for DOD.

From the aforementioned history related to the acquisition authority for UAS programs and the current magnitude of effort, responsibility and funding towards UAS programs, it is imperative to make the right decision now. The economic savings associated with combining similar and redundant UAS programs are at stake, and the establishment of a sound acquisition baseline for future UAS programs that advance the current strike and ISR mission. Ramifications from not addressing this issue in the unmanned aerial domain could cause unneeded duplication and fiscal waste in future unmanned ground and maritime based programs as they come into similar maturity.

Definitions

BAMS: Broad Area Maritime Surveillance--A high altitude, long endurance ISR UAS being developed by the US Navy. It is a program to develop a multiple-sensor, persistent maritime ISR UAS that provides persistent ISR to supported commanders. BAMS UAS will be a force multiplier for the Joint Forces and fleet commanders: it will enhance the situational awareness of the battle space and shorten the sensor-to-shooter kill chain. The BAMS UAS will provide DoD with a unique capability to persistently detect, classify, and identify maritime targets within a large volume of the maritime battle space. IOC is planned for 2015.[26]

Economy of scale: A term used by economists to refer to the situation in which the cost of producing and additional unit of output of a product decreases as the volume of output increases.[27]

MQ-1B Predator: A medium altitude, armed UAS employed by the United States Air Force (USAF). The MQ-1 Predator began in 1994 and transitioned to the Air Force in 1997. Since 1995, Predator has flown surveillance missions over Iraq, Bosnia, Kosovo, and Afghanistan. It is armed with two AGM-114 Hellfire missiles and includes full-motion electro-optical (EO) and infrared (IR) sensors as well as synthetic aperture radar (SAR) capabilities.[28]

MQ-1C Sky Warrior: A medium altitude, armed UAS employed by the US Army formerly known as extended range/multipurpose (ER/MP). The Sky Warrior's payload includes full motion EO/IR sensors with synthetic aperture radar moving target indicator (SAR/MTI) capabilities. Additionally, two 250-pound and two 500-pound hard points under the main wings provide an attack capability.[29]

MQ-9 Reaper: Medium altitude, heavily armed UAS employed by the USAF. The MQ-9 is a medium- to high-altitude, long-endurance UAS. Its primary mission is to act as a persistent hunter-killer for critical time-sensitive targets and secondarily to act as an intelligence collection asset. The integrated sensor suite includes a SAR/MTI capability and a turret containing full motion, electro-optical and mid-wave IR sensors, a laser rangefinder, and a laser target designator. The Reaper can carry 3,000 pounds of external payload.

Procurement: Funding for procurement buys new weapons and other equipment that DOD needs to carry out its missions in peacetime and to prepare for war. The funds cover a wide array of items ranging from aircraft, ships, missiles, automobiles and related military support equipment.[30]

RDT&E: Research, Development, Test and Evaluation pay for basic and applied research, fabrication of devices for demonstrating new technologies, development and testing of prototypes, and testing of full-scale models of weapon systems before they enter production. Development funds also pay for operational testing of systems, when they are first taken into the field and when they are modified during the course of operations.[31]

RQ-4 Global Hawk: An Air Force high-altitude, long-endurance unmanned aircraft designed to provide wide area coverage of up to 40,000 square nautical miles per day. It's EO/IR and SAR/MTI sensors allow day/night, all-weather reconnaissance. The first operational production aircraft, the Block 10 –A model," deployed in January 2006 to U.S. Central Command (CENTCOM) and replaced the prototype ACTD configuration, which had been deployed there for most of the time since 2001.[32]

UAS: Unmanned Aircraft System--A powered vehicle that does not carry a human operator, can be operated autonomously or remotely, can be expendable or recoverable, and can carry a lethal or nonlethal payload. UAS are combination of equipment required to operate the UAS capability to include the aerial vehicle with payload, ground control station (GCS), launch and recovery element (LRE), and associated communication systems.[33]

Limitations

Three limitations apply to this thesis. First, planned funding only exists up to FY13 per the President's annual budget request. Any funding comparisons beyond this time frame are purely speculative. Secondly, the study worked within the current market production capacity for UAS airframes and related equipment to arrive at realistic funding efficiencies. Surge production capability for contractors could not be obtained. Lastly, the total cost and total quantity of BAMS aircraft are still undetermined. The first BAMS aircraft is not scheduled for procurement until 2014, which is beyond the current FYDP.[34]

Delimitations

During this study, several issues were intentionally not addressed. First, due to the inherent similarities and potential for consolidation of acquisition programs, this thesis only analyzed the following medium and high altitude UAS systems: MQ-1B Predator (USAF), MQ-1C Sky Warrior (USA), MQ-9 Reaper (USAF), RQ-4 Global Hawk (USAF), and the BAMS (USN). It does not include small and low altitude UAS in the analysis of DOD benefits and drawbacks as they do not represent a majority of the DOD funding.

Second, only funding related to research, development, test and evaluation (RDT&E) and procurement were considered in this study. Operations and Maintenance (O&M) and Military Construction (MILCON) were purposefully excluded as they do not directly reflect sound acquisition decisions of the best UAS platform choice. O&M and MILCON do have second and third order funding and efficiency implications related to UAS operations such as training, supplies, maintenance, and hangars 35

Third, this thesis avoided command and control issues with respect to UAS. The issue of operational control of UAS in theater with respect to ISR priority was not considered. The argument between the Army's organic employment methods versus the Air Force's remote operation model has also been excluded. While, both of these methods have funding implications for UAS systems, but do not directly impact the acquisition aspect of the program. Additionally, rated officer versus enlisted pilots of UAS did not contribute to this study from an acquisition perspective and has been excluded as well.

Although joint basing offered several benefits pertaining to common maintenance, manpower and civilian airspace deconfliction, these topics did not directly apply to the acquirement of the best UAS to fulfill a stated DOD requirement. Joint basing also brings with it the complicated issue of Congressional involvement when moving missions and personnel from one district to another in the US. Additionally, joint basing in combat zones does not affect research and procurement as the UAS have already been fielded.

Similarly, Congressional funding changes to the DOD budget were not taken into consideration as partisan, state and local agendas did not necessarily reflect the best acquisition direction for the DOD. Only the fiscal data contained in the President's Budget was analyzed as this was the Executive Branch's final decision on the proper force structure and direction for UAS development, although it is recognized that Congress provides the final authorization and appropriation of all DOD funding.

Finally, this thesis does not include the medium and high altitude UAS programs of other US Departments. The Department of Homeland Security (DHS), National Oceanic and Atmospheric Administration (NOAA) under the Department of Commerce, and the National Aeronautics and Space Administration (NASA) all have medium and high altitude UAS projects to address respective requirements. Although the above departments utilized variants of the same medium and high altitude UAS in this study, the capability and mission requirements varies greatly based on the mission of each organization.

Assumptions

First, individual services will accept the analysis and programming decision made by the Office of the Secretary of Defense (OSD) and comply with any restructuring measures associated with a single entity acquisition authority for the selection of medium and high altitude UAS programs. Secondly, there will be no departmental decision changing the acquisition authorities within the DOD for the duration of this study. Additionally, no new medium or high altitude UAS will be entering the DOD inventory similar to the selected five UAS. Next, consolidation or merging of UAS platforms will reduce unit costs based on the economies of scale concept. Finally, the long-term economic efficiencies and interoperability benefits will outweigh short-term second and

third order effects of merging UAS systems. Second and third order changes affecting

personnel training, infrastructure, support and operational employment, to name a few,

should generate short-term expenses for DOD and the respective services until UAS

program consolidations are complete.

Summary

Medium and high altitude UAS programs have existed for decades, but recent

developments have raised concerns on funding inefficiencies and potential duplication of

capabilities that this thesis will address. This chapter provided a brief history on the

introduction of medium altitude UAS programs in the DOD as well as the significance of

this topic. Chapter 2 will provide an overview of the sources of literature utilized in this

study and their importance to the thesis.

---

[1]Fox News, "Obama Calling for Defense Budget Cuts," http://www.foxnews.com/ politics/2009/01/30 (accessed 21 April 2009).

[2]Office of the Secratary of Defense, *Unmanned Systems Roadmap 2007-2032* (Washington, DC: Government Printing Office, 2007), 10.

[3]Government Accountability Office. *GAO/T-NSIAD-97-138 DOD UAV Acquisition Efforts* (Washington, DC: Government Printing Office, 1997), 2.

[4]Ibid.

[5]Ibid., 2-3.

[6]US Air Force Briefing, 6 February 2008, *USAF Unmanned Aircraft Systems Vision* (Washington, DC: AF/A2, 2008), 4.

[7]Government Accountability Office. *GAO-04-342 Force Structure: Improved Strategic Planning Can Enhance DOD's UAV efforts* (Washington, DC: Government Printing Office, 2004), Introduction.

[8]Government Accountability Office, Changes in Global Hawk's Acquisition Strategy Are Needed to Reduce Program Risks, http://www.gao.gov/products/GAO-05-6, (accessed 30 April 2009).

[9]Tactical Intelligence Systems Directorate, *DOD Joint UAV Master Plan – 1988*, (Washington, DC: OASD-C3I, 1988), 1.

[10]Government Accountability Office. *GAO-04-342 Force Structure: Improved Strategic Planning Can Enhance DOD's UAV efforts* (Washington, DC: Government Printing Office, 2004), 6-7

[11]Ibid.

[12]Government Accountability Office. *GAO-04-342 Force Structure: Improved Strategic Planning Can Enhance DOD's UAV efforts* (Washington, DC: Government Printing Office, 2004), 6-7.

[13]Ibid.

[14]Ibid.

[15]John Young, Memorandum on *Unmanned Aircraft Systems* (Washington, DC: OSD AT&L, 2007)

[16]US Congressional Armed Service Committee Conference Report, *Floyd D. Spencer National Defense Authorization Act*, Washington, DC: US Congress, 2001, Section 220.

[17]US Congressional Armed Service Committee Conference Report, *John Warner National Defense Authorization Act*, Washington, DC: US Congress, 2006, Section 141.

[18]Ibid., 14.

[19]Ibid.

[20]Department of Defense, ―Contracts," http://www.defenselink.mil/contracts/contract.aspx?contractid=3758; (accessed 11 February 2009).

[21]Northrop Grumman, ―Broad Area Maritime Surveillance (BAMS)," http://www.irconnect.com/noc/press/pages/news_releases.html?d=140693 (accessed 2009).

[22]Office of the Secretary of Defense, *Fiscal Year 2009 Budget Request SummaryJustification* (Washington, DC: Government Printing Office, 2008), 163.

[23]AM General, ―HMMWV Background," http://www.amgeneral.com/vehicles/hmmwv/background.php (accessed 1 May 2009).

[24]Ibid.

[25]Linux Project, Economies of Scale Definition, http://www.linfo.org/economies_of_scale.html (accessed 1 May 2009).

[26]Ibid.,75.

[27]Linux Project, Economies of Scale Definition, http://www.linfo.org/economies_of_scale.html (accessed 1 May 2009).

[28]Ibid., 65.

[29]Ibid., 66.

[30]Congressional Budget Office, CBO's Estimate of a Sustaining Budget for National Defense, http://www.cbo.gov/doc.cfm?index=2398&type=0&sequence=3. (accessed 23 April 2009).

[31]Ibid.

[32]Ibid., 68.

[33]Ibid., 1.

[34]Deagal, RQ-4 Global Hawk, http://www.deagel.com/AEWandC-ISR-and-EW-Aircraft/RQ-4A-Global-Hawk_a000556001.aspx (accessed 10 May 2009).

[35]Defense Acquisition University, ―Operations and Maintenance Funds," https://acc.dau.mil/CommunityBrowser.aspx?id=28980 (accessed 4 March 2009).

CHAPTER 2

LITERATURE REVIEW

In current literature related to the acquisition of medium and high altitude UAS, two prominent themes emerge. The first theme is the acquisition process and inherent inefficiencies of DOD policy concerning UAS. The second is the operational requirements of current medium and high altitude UAS in the current and future force structure. This chapter is focused on the review of these documents and their application to the research questions associated with this study.

Acquisition

In 2002, the Joint UAS Planning Task Force created the *UAV Roadmap 2002-2027* in an attempt to address long-term procurement objectives and funding requirements. This document attempted to address the entire UAS program within the DOD but fell short in two key areas: lack of a DOD strategic plan concerning UAS programs and the authority to mandate UAS portfolio changes in the best interest of the DOD.[1] It was from this document that DOD received most of its criticism for poor acquisition practices and direction. At this point in time, the true impact of UAS on the battlefield had not yet matured. MQ-1 Predators and the test version of the RQ-4 Global Hawk were the only two of the five UAS contributing to OEF. The MQ-1C Sky Warrior, MQ-9 Reaper and BAMS had yet to become operational. Although this roadmap lacked a strategic plan, many of the key UAS capabilities were still theoretical at this point in time.

A March 2004 GAO report on the *UAV Roadmap 2002-2027* criticized two shortfall areas; a lack of DOD strategic planning concerning UAS programs and the authority to mandate UAS portfolio changes in the best interest of the DOD.[2] The concern was that if left unchecked and uncoordinated, the acquisition programs of each service would continue on divergent paths and lead to force structure problems and integration issues once these platforms were fielded. As noted in the report,

> Without a strategic plan and an oversight body with sufficient authority to implement the plan, DOD has little assurance that its investment in UAVs will be effectively integrated into the force structure. Consequently, DOD risks poorly integrating UAVs into the force structure, which could increase development, procurement, and logistics costs, and increase the risk of future interoperability problems.[3]

> Our concern, however, is that neither the Roadmap nor other defense planning documents represent a comprehensive strategic plan to ensure that the services and other DOD agencies focus development efforts on systems that complement each other, will perform the range of priority missions needed, and avoid duplication.[4]

Without strategic DOD direction, the services are not bound to a common strategic direction. This led to the possibility of multiple service specific UAS platforms conducting similar missions, but procured under different contracts. Without the establishment of a strategic goal for numbers, types and missions of UAS, inefficiency and duplication are bound to occur. Supporting and integrating all of these disparate UAS in future combat environments could prove challenging.

Anticipating the myriad of potential problems realized in future years, the GAO report and testimony provided several solutions to Congress highlighting doctrinal and organizational recommendations. The most pointed of these solutions called for one entity to hold oversight and fiscal authority of all UAS programs and designate the UAV Task Force or another appropriate organization to oversee the strategic plan's implementation, providing it with sufficient authority to effectively enforce the plan's direction, and promote joint operations and efficient expenditure of funds.[5]

Although this report offered a potential solution, DOD disagreed with the above recommendation. OSD AT&L stated it did not need to provide additional authority to an organization within the department. DOD believed that OSD AT&L had the appropriate oversight and influence to integrate UAS capability to combatant command (COCOM) operational forces. Department leadership also believed that through OSD, the annual Planning, Programming, Budgeting, and Execution (PPBE) process and the Joint Capabilities Integration and Development System (JCIDS), provided the opportunity to adequately review and enforce program activities, including UAV activities, across Services."[6] The PPBE process followed SECDEF established goals, policy and strategy in the allocation of DOD resources. The JCIDS assessed current and future capabilities to allow joint forces to meet emerging military challenges.[7]

Although very factual in response and authoritative from a departmental perspective, the fact remained that there was no champion solely for UAS programs with the above mentioned authority. This lack of program directive authority made the DOD UAS Planning Task Force only an advisory body. The respective services still maintained funding authority over all of their UAS systems in RDT&E, procurement and fielding, regardless of how similar the platform.[8] The DOD UAS Planning Task Force could not compel any service to adopt or implement any programmatic or production recommendations, no matter how beneficial it was to the defense force structure. Chapter 4 will determine if this lack of authority allowed the perpetuation of DOD funding inefficiencies.

In 2005, the DOD strategic UAS plan still lacked service integration and coordination within the department. The ramifications of this disconnect allowed UAS development and investment efforts to continue without authoritative oversight focused solely on UAS. The GAO report Improved Strategic and Acquisition Planning Can Help Address Emerging Challenges targeted the DOD coordination and integration shortfalls:

> DOD's UAV Roadmap contains some elements of a strategic plan, but it does not describe the interrelationship of service roadmaps to the DOD Roadmap or clearly identify funding priorities. Thus, DOD may not be well positioned to make sound program decisions or establish funding priorities, nor will Congress have all the information it needs to evaluate funding requests. Such a plan would also help DOD minimize the types of challenges that are emerging.9

> For example, the Roadmap did not include a mission statement, description of how program evaluations were used to establish or revise goals, discussion of the interrelationship between service plans and programs to develop and field UAVs, or provide adequate information on current and projected funding needs. In particular, it does not explain the interrelationship between service-specific efforts, identify opportunities for joint endeavors, or address funding issues.[10]

The lack of control and effective management are at the crux of the efficiency issue pertaining to the acquisition of DOD UAS medium and high altitude platforms. Without an organization capable of understanding and directing the research, development, procurement and fielding of expensive UAS system across the DOD, funding inefficiencies and poor acquisition decisions were bound to occur unabated. The lack of a mission statement from a DOD level organization and lack of evaluation criteria provided the respective services neither direction nor motivation to collaborate towards a common departmental goal. In the absence of a higher level mission statement, each service conducted acquisition practices that were in their own best interest.

A prime example of independent acquisition practice occurred with the Air Force RQ-4 Global Hawk and Navy BAMS UAS in 2008. A single contract would have saved significant funding, but that entailed appointing one organization in charge. But the single acquisition recommendation concerning the Global Hawk and BAMS programs was not a new concept for the DOD in 2008. Back in 2004, the Defense Science Board (DSB) study on *Unmanned Aerial Vehicles and Uninhabited Combat Aerial Vehicles* came to the exact same conclusion. It recommended that DOD ―Merge the current USAF Global Hawk and Navy Broad Area Maritime Surveillance (BAMS) programs into a common use High Altitude Endurance (HAE) UAV system that will meet the needs of both Services"[11] Four years before the BAMS contract ever became official, the DSB stated that the DOD should ―modernize and transform the business of defense, getting the best value for the taxpayer's money." The DSB had already come to the conclusion that a single acquisition authority would be the best route to bring both weapon systems into the DOD force structure.[12] Yet in 2008, the DOD possessed two services with two contracts for the same airframe.

In November 2008, the GAO released another report to Congress titled *UAS: Additional Actions Needed to Improve Management and Integration of DOD Efforts to Support Warfighter Needs.* It focused on the analysis of the newly released *DOD UAS Roadmap 2007-2032.* This report uncovered UAS management and integration problems and proposed organizational, visionary and planning actions needed to correct various inefficiencies that continue to plague the development of UAS. Serious concern stemmed from the exponential growth of UAS systems in military inventories and the problems that accompanied such an expanding mission. This report recommended to the House Armed Services Committee (HASC) that the Secretary of Defense:

designate a single departmental entity responsible and accountable for integrating all cross-cutting DOD efforts related to UAS" and ~~develop~~ a UAS strategic plan to align and integrate respective departmental and service efforts and funding with long-term goals.[13]

As evident in the 2004 and 2005 GAO testimonies and reports, very similar concerns and recommended actions were proposed for DOD implementation. DOD officially did not concur with either of these two recommendations made to the House Armed Service Subcommittee on Air and Land Forces. DOD commented on the single responsible entity recommendation that ~~it~~ had created the UAS Task Force--in lieu of an executive agent--to coordinate critical UAS issues to enhance operations, enable interdependencies, and streamline UAS acquisition."[14]

This DOD response appeared to quell the GAO recommendation, but the newly created UAS Task Force in the Office of the Secretary of Defense only coordinated and enabled. The Defense Department distributed accountability for six UAS identified issue areas among multiple organizations.[15] Acquisition, research and development, standardization and interoperability, civil airspace, payload and sensor integration and bandwidth utilization issues were to be managed by OSD, the three service departments, and Joint Staff offices. Leadership of each issue area was assigned by the organization with the most applicable expertise.[16] For example, the Airspace Integration Integrated Process Team (IPT) was co-chaired by the OSD AT&L and Air Force. The Payload and Sensor Integration IPT was co-chaired by the OSD AT&L and Army.[17] But with no single authority in control of funding, coordination and enabling efforts only worked if all service stakeholders agreed to implement recommendations made by the other issue groups.

DOD did not concur with the second GAO recommendation to develop a strategic plan to integrate service efforts and funding with long term goals.

> DOD stated that it has undertaken several initiatives to improve the department's approach to investment and decision making, including the implementation of its capability portfolio managers, and that the department's strategic plan for investment is aligned with portfolios that address specific warfighting capabilities as opposed to platforms or material solutions, such as UAS. DOD also stated that long-term goals and guidance for achieving those goals are provided in top-level documents, such as the Guidance for the Development of the Force, and that the Joint Capabilities Integration Development System provides a structured process to address warfighting capability and capacity gaps.[18]

Although there are several top level DOD documents such as DOD 5000 and the Joint Capabilities Integration and Development System JCIDS that provided strategic guidance on force development and joint capability integration of numerous platforms, these did not substitute as a strategic plan to integrate UAS efforts and funding to accomplish long-term UAS goals. In the case of the *DOD Unmanned Systems Roadmap 2007-2032*, numerous exemptions lead to questions on the true direction OSD intended for unmanned aerial systems. This roadmap lacked the measures required to accomplish UAS objectives, track progress and performance gaps and correlate UAS investments with long-term goals.[19] In the UAS Task Force's defense, without control of the funding for UAS systems, service acquisition priority changes could cause severe disruption to the long-term goals of such a higher level coordination organization. This report highlighted the importance and benefits of identifying one single departmental entity responsible, accountable and in control of funding to efficiently progress the advancement of UAS platforms.

In November 2007, DOD Directive 5000.1 provided the overarching policies and principles behind the direction of defense procurement. This document established five

25

policies for defense acquisition. The policies of flexibility, responsiveness, innovation, discipline and streamlined and effective management were utilized as desired procurement practices in this thesis.[20] The recommendations of this study attempted to single out DOD acquisition practices that did not meet the spirit of the DEPSECDEF's intent. These five policies were paramount to the accomplishment of this thesis. Without a defense department ideal of what procurement should entail, the topic becomes very subjective. But DOD 5000.1 provided the validity behind the analysis, conclusions and recommendations of this study.

Fiscal data from the FY09 President's Budget, the Office of the Secretary of Defense Comptroller's Fiscal Year 2009 Budget Request Summary and the UAS TF were utilized in the fiscal analysis of this thesis. These budget documents released in 2009 compiled funding amounts by major investment catagories for the selected UAS acquisition programs.[21] These budget documents reflected the DOD's programmatic intent for all programs, prior to changes instituted by Congress. The seperation of funding catagories allowed the five identified UAS programs to be isolated into different funding catagories in order to understand the RDT&E and procurement decisions and amounts. This allowed for the magnitude of the five selected UAS to be visualized through analysis. The UAS TF funding data was a source that had the oversight and funding detail neccesary to identify the specifics associated with UAS from a variety of historical perspectives. By understanding the scale involved, this study could not only show the investment focus for DOD UAS, but also relays the importance of effectively managing the large funding amounts associated with the various UAS.

In order to determine scale, the Department of Defense's approved President's Budget for all unmanned categories from FY07 – FY13 was referenced in this study. In table 1, of the $24.3 billion programmed for all unmanned systems, an overwhelming $21.6 billon was dedicated solely to aerial systems. To contrast this point, figure 2 provided a visual depiction on the relative amounts of funding between UAS and the other two genres; unmanned ground vehicles (UGV) and unmanned maritime systems (UMS). With nearly 89 percent of the DOD unmanned portfolio, UAS offered the largest potential for significant cost savings based on scale and visibly increasing budget.

Table 1.   FY07-13 President's Budget for Unmanned Systems

PORs FY08PB ($M)	Funding Source	FY07	FY08	FY09	FY10	FY11	FY12	FY13	TOTAL
UGV	RDT&E*	$198.2	$215.4	$199.8	$167.5	$129.3	$58.5	$20.0	$989
	PROC*	$106.5	$39.3	$29.7	$18.3	$17.9	$156.3	$481.5	$849
	O&M*	$156.0	$5.7	$8.8	$10.3	$11.0	$12.1	$12.7	$217
UAS	RDT&E	$760.8	$814.8	$1246.7	$1334.9	$1181.8	$859.1	$839.5	$7,038
	PROC	$878.4	$1370.3	$2025.1	$2010.5	$1725.7	$1750.8	$1585.7	$11,346
	O&M	$590.0	$352.3	$367.7	$421.2	$458.8	$501.5	$552.0	$3,244
UMS	RDT&E	$43.8	$22.7	$34.5	$77.0	$86.0	$101.9	$131.9	$498
	PROC	$1.7	$2.7	$3.2	$4.8	$40.8	$25.0	$25.1	$103
	O&M	$4.3	$3.1	$2.8	$2.3	$3.9	$5.9	$6.9	$29
TOTAL		$2731.5	$2825.4	$3949.6	$4041.6	$3657.3	$3461.3	$3643.5	$24,310

* RDT&E = Research, Development, Test, and Evaluation; PROC = Procurement; O&M = Operations and Maintenance

Source: Office of the Secratary of Defense, *Unmanned Systems Roadmap 2007-2032*, (Washington DC: Department of Defense, 2007), 10.

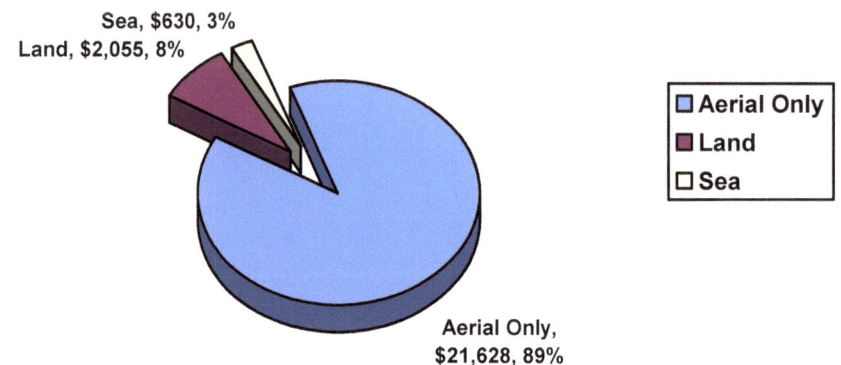

Figure 2.   Total DoD Unmanned Funding over the FY07-FY13 FYDP
Source: Office of the Secratary of Defense, *Unmanned Systems Roadmap 2007-2032*, (Washington DC: DOD, 2007), 10.

Now compare the 1988 total for UAV research and procurement amount of $105.9M,[22] with the 2009 total for UAS research and procurement amount of $3271.8M in figure 3.[23] This represented a 3,086 percent increase in Defense wide UAS funding. From a fiscally responsible position, previous inefficiencies that wasted a few million now could become several hundred million wasted due to the same poor acquisition practices. The levels of funding involved are too large to continue with the status quo acquisition methods of the past. UAS can no longer be viewed as small, separate research projects. These systems must be considered as major weapons systems would benefit from concepts such as economies of scale to reduce costs.

Figure 3.   UAS RDT&E and Procurement Funding 1988 – 2013
Source: DOD UAS Task Force briefing 9 January 2009, Unmanned Aircraft Systems
(Washington DC: Government Printing Office, 2009), 4.

Finally, the respective organizations involved with DOD acquisition and UAS

issues were researched to identify the exact role and mission they respectively played in

the DOD organizational structure. The Congressional Budget Office (CBO), Defense

Acquisition University (DAU), DSB, GAO, OSD AT&L and UAS TF identified the

responsibilities and fiscal authorities of each organization the exact link where oversight

and funding authority began. Understanding the interrelationship and authority of each

organization allowed for accurate recommendations to be made pertaining to the DOD

acquisition process. The purview and relationship to the DOD budget and acquisition

process also determined the respective credibility of each source.

Requirements

On 6 April 2006, the Government Accountability Office delivered testimony to the HASC Subcommittee on Tactical Air and Land Forces on *Improved planning and acquisition strategies can help address operational challenges.*[24] This testimony reviewed recent actions by DOD to address UAS management and requirement shortfalls. Despite the creation of several new DOD UAS focused organizations; the GAO identified several operational challenges.

> While DOD intends for these entities to play a role in guiding service UAS acquisition, planning, prioritization, and execution of unmanned air systems, it is also unclear to what extent they will be able to influence the services because none of the entities are chartered with the authority to direct the military services to adopt any of their suggestions. Rather, they act in an advisory capacity and make recommendations to the services and Joint Requirements Oversight Council.[25]

Again, the authority to direct change is still the prime issue surrounding UAS acquisition according to the GAO. This testimony did follow the recent creation of the JUAS CoE and MRB, and no examples of successful implementation were mentioned.

In August 2008, the C4ISR Journal included an article focused on UAS ISR issues titled Who Should Fly by Ben Iannotta? An issue raised by Mr. Iannotta addressed the weaponization the US Army planned for its MQ-1C Sky Warrior fleet. Up to this point, the US Army has primarily focused its UAS efforts toward reconnaissance and to provide situational awareness to ground forces in fire fights.26 Seeing the benefits of near-real time intelligence and near-real time weapons delivery utilized by the US Air Force, the Army moved to arm future UAS. The Army then began to field the MQ-1C Sky Warrior UAV, an MQ-1B Predator derivative to replace the older and limited MQ-5 Hunter UAV.27 This article provided operational level views of the Army and Air Force on UAV development, but did not address future force structure ramifications. As the services evolve to improve tactics, weapons, and aircraft to address the current counter-insurgency (COIN) threat, there are serious ramifications that result from acquisition efforts conducted without synchronization. The concerns raised by separate procurement of very similar aircraft and armament lead to potential economic, integration and maintenance inefficiencies that should concern the DOD.

A January 2009 Defense News interview with Lt Gen David Deptula discussed many topics surrounding the current and future plans for UAS in the Air Force and DOD. According to Lt Gen Deptula, the Air Force Deputy Chief of Staff for ISR, one critical challenge facing UAS is ―the streamlined and efficient acquisition of our ISR systems.‖28 The Obama administration plans to increase non-discretionary funding in areas like Social Security and Medicare. This will lead to increased scrutiny of discretionary funding portfolios like Defense.29 This interview provided insight from a service intelligence chief. Deptula offered that if efforts are not made to optimize UAS platforms, funding decisions would be made from higher that could impact defense capabilities. He understands that DOD must optimize its capabilities and resources in order to maintain ISR superiority. Starting the analysis now for ISR efficiencies would allow the DOD to manage its fleet of UAS in a way that ensured no degradation of current capability. Lt Gen Deptula specifically mentioned the RQ-4 Global Hawk and BAMS UAS programs as a potential model for joint acquisition.

Understanding the duplication of capability concerns, two key sources were utilized to determine the level of similarity of the five selected UAS platforms. First, the DOD UAS Roadmap 2007-2032 provided DOD's most current policy on UAS requirements and desired capabilities. This document included the strategic view for

UAS procurement, integration considerations and flight performance and payload

consideration background on every UAS system in the DOD inventory. Most capability

and payload data in this study was derived from the UAS Roadmap 2007-2032.

Secondly, contractor data from General Atomics and Northrop Grumman was

compared to the DOD statistics to determine any gross differences in stated flight

performance and payload capability levels of the five UAS in this study. Fact sheets on

their respective UAS were utilized to confirm contractor claims of performance.

Compared with DOD capability information from the UAS Roadmap 2007-2032, these

two sources provided data that was cross-referenced to identify any gross differences in

stated capability. Utilizing two independent sets of performance statistics reduced the

impact of inaccurate or biased claims. If the numbers were similar they validated one

another. If the numbers varied greatly, the associated analysis had to reflect these

caveats. When there was a significant disparity between numbers, the DOD data was

given priority, but complementary analysis was still conducted with the contractor data

and the results were displayed to reflect the implications.

Summary

Several sources comprised the literature review, but they addressed two main

areas, acquisition and requirements. There appears to be a mounting case to appoint a

single authority to mitigate the economic inefficiencies associated with the procurement

of medium and high altitude UAS. The literature also identifies concerns with

duplicative capabilities resident in current UAS programs. It is from these two trends that Chapter 3 will outline the associated research methodology of this thesis and explain

the rationale for its construct.

---

[1]Government Accountability Office. *GAO-04-342 Force Structure: Improved Strategic Planning Can Enhance DOD's UAV efforts* (Washington, DC: Government Printing Office, 2004), Introduction.

[2]Ibid.

[3]Ibid.

[4]Government Accountability Office. *GAO-04-530T UAVs: Major management issues facing DOD's development and fielding efforts* (Washington, DC: Government Printing Office, 2004), 2.

[5]Ibid., 3.

[6]Glenn Lamartin, *USD AT&L Response to Draft Report GAO-04-342 FORCE STRUCTURE* (Washington, DC: DOD, 5 March 2004).

[7]Defense Acquisition Guidebook, Joint Capabilities Integration and Development System, https://akss.dau.mil/dag/DoD5000.asp?view=functional (accessed 23 April 2009).

[8] Ibid.

[9]Government Accountability Office. *GAO-05-395T Improved Strategic and Acquisition Planning Can Help Address Emerging Challenges* (Washington, DC: Government Printing Office, 2005), i.

[10]Ibid., 3.

[11]Defense Science Board, *Unmanned Aerial Vehicles and Uninhabited Combat Aerial Vehicles* (Washington, DC: Government Printing Office, 2004), vi.

[12]Defense Science Board, ─About the DSB," http://www.acq.osd.mil/dsb/ (accessed 1 March 2009).

[13]Government Accountability Office. *GAO-09-175 Additional Actions Needed to Improve Management and Integration of DOD Efforts to Support Warfighter Needs* (Washington, DC: Government Printing Office, 2008.), 5-6.

[14]Ibid., 28-29.

[15]Ibid., 29.

[16]Ibid., 14.

[17]DOD UAS Task Force briefing 29 November 2007, Unmanned Aircraft Systems Task Force, (Washington, DC: Government Printing Office, 2009), 3.

[18]Ibid., 30.

[19]Ibid., 31.

[20]Department of Defense, DOD 5000.01, *The Defense Acquisition System* (Washington, DC: Government Printing Office, 200), 3.

[21]Office of the Secretary of Defense. *Fiscal Year 2009 Budget Request Summary Justification* (Washington, DC: Department of Defense, 2008).

[22]Tactical Intelligence Systems Directorate, *DOD Joint UAV Master Plan – 1988* (Washington, DC: OASD-C3I, 1988), 45-48.

[23]Office of the Secratary of Defense, *Unmanned Systems Roadmap 2007-2032* (Washington, DC: DOD, 2007), 10.

[24]Government Accountability Office. *GAO-06-610T Improved planning and acquisition strategies can help address operational challenges* (Washington, DC: Government Printing Offfice, 2006), 13-14.

[25]Ibid.

[26]Iannotta, Ben, ―Who should fly?" *C4ISR Journal*, August 2008: 14-16.

[27]Ibid.

[28]Defense News, "Lt Gen David Deptula: US Air Force Deputy Chief of Staff for ISR," http://www.defensenews.com/story.php?i=3896251 (accessed 12 January 2009).

[29]Ibid.

CHAPTER 3

RESEARCH METHODOLOGY

The purpose of this study is to examine the economic efficiencies gained from the appointment of a lead service or agency to direct the procurement of the five UAS systems under review: MQ1-B Predator (USAF), MQ-1C Sky Warrior (Army), MQ-9 Reaper (USAF), RQ-4 Global Hawk (USAF), and the BAMS (USN). The general research approach identifies the amounts of funding planned for each of the systems and their respective capabilities.

This chapter is divided into four sections. The first section identifies the steps taken to obtain information related to this thesis, while the second frames the research criteria utilized in this thesis. Next, applied research methodology followed and depicts the identified areas of analysis. The fourth section identifies strengths and weaknesses related to this study.

Research Criteria

The validity of economic benefits associated with the appointment of a lead acquisition agency required proving two points; the existence of funding inefficiencies and duplication of operational capabilities. In order to assess funding inefficiencies and operational requirements, the five UASs are divided into two distinct groups. Based on similar mission profiles, the MQ-1B Predator, MQ-1C Sky Warrior, and the MQ-9 Reaper represent one comparative grouping. This group will be referred to as the Medium Altitude UAS group. The second group is comprised of the RQ-4 Global Hawk and BAMS which will be referred to as the High Altitude UAS group.

The rationale behind the Medium Altitude group of UASs stems from several similarities. All three UAS systems are designed, manufactured and delivered by the same company, General Atomics.[1] Accordingly, all three airframes are built and constructed side-by-side in various facilities in the San Diego area. Secondly, all three of the medium -altitude UASs are armed aerial vehicles. While all these unmanned aircraft employ the AGM-114 Hellfire air-to-surface missile, the MQ-9 Reaper is also capable of carrying a more diverse payload of weapons due to airframe capacity. Finally, all three medium-altitude UASs employ similar types of intelligence collection assets in the form of electro-optical (EO), infrared (IR), synthetic aperture radar (SAR) and various communications intelligence payloads.

The rationale for selecting the high altitude UAS group is very similar to that of the medium altitude UAS group. Much like the medium altitude UAS group, the Air Force RQ-4 Global Hawk and the Navy's Broad Area Maritime Surveillance are produced by the same company, Northrop Grumman.[2] Like their medium altitude counterparts, both employ EO/IR, SAR and various communication intelligence payloads.[3] Both of these high-altitude UASs operate at altitudes in the 60,000ft range in order to maximize the area of coverage as they collect information over vast expanses of land and water.[4]

From the selected UAS, research criteria had to identify their significance pertaining to economic inefficiency and associate each with UAS capability duplication. If the scale of economic significance proved minuscule or if the selected UAS possessed little redundancy, the justification for a single acquisition authority would be less convincing. The key research criteria associated with economic inefficiency are threefold. First, the magnitude of UAS investments compared to all unmanned investments indicates the current focus of the DOD. Secondly, the proportion of the entire DOD UAS budget that the five selected UAS comprise indicates the focus of service efforts in order to meet requirements. Finally the trend of medium and high altitude UAS investments in the past and near future indicates linear or exponential changes and the ramifications associated with the selected UAS. This study's assessments will reflect the existence of any inefficiency that could be mitigated by the appointment of an acquisition authority.

The key research criteria associated with duplication of capability are threefold. The first criteria will base UAS comparisons on performance: maximum altitude, speed, endurance, radius and payload. The second will evaluate payload integration followed by operational risk. Qualitative and quantitative research will compare these criteria against all five selected UAS.

Research Methodology for Primary and Secondary Questions

In order to discern if a single UAS authority is beneficial to the DOD, analysis must center on the funding magnitude and system capabilities. The funding magnitude will determine if there is enough investment and inefficiency in these UAS programs to warrant introducing a cost saving authority at the DOD level. Evidence of both is mandatory for recommending programmatic changes. Secondly, analysis of system capabilities will determine if redundancy exists between UAS platforms that could be reduced under the direction of a single DOD authority.

The first approach centers on the funding amounts associated with all five UAS programs. Due to the investment nature of these funds in relation to the acquisition process, the only types of funding considered are the research, development, test and evaluation (RDT&E) and procurement. Both are required to build new weapon systems. This study intentionally left out operations and maintenance (O&M) funds as these are utilized for day-to-day operations. While O&M funds cover activities related to military operations, civilian pay, training, and sustainment, they do not necessarily reflect good or bad acquisition practices. [5]

The research criteria for funding incorporated two distinct analytical efforts. First, research will attempt to identify a significant funding growth in UAS investments from a historical perspective. This will attempt to convey the rapid fiscal increase involved with medium and high altitude UAS. Secondly, this research will discern the percentage of the entire UAS portfolio that these five systems account for. By utilizing a common year of reference, analysis can be conducted to determine if these five UAS account for a majority or minority of all DOD UAS investments.

The second approach centers on the military capabilities of the five UAS programs. In the design and acquisition of weapon programs, similar requirements and capabilities can lead to waste and duplication of effort. In order to determine if a lead agency would be beneficial, duplication of capability must be identified in the current UAS programs. The more UAS programs exhibiting redundant capabilities strengthen the argument for one organization to possess the authority to mandate change.

Analysis in this study focused on two key UAS components. First, airframe flight performance was quantitatively analyzed to identify similar capabilities. The five UAS are compared in five separate performance areas: maximum altitude, speed, endurance, radius and payload. A similarity value of 75 percent was assigned as an indicator of potential consolidation of mission requirements between air vehicles. This percentage level reflected enough commonality in performance to warrant a more detailed analysis of the required capability between weapons systems.

This study chose a higher 75 percent commonality threshold to narrow the results to those UAS pairs that exhibit the greatest probability for program consolidation. The 75 percent value is based on the Army's Future Combat System (FCS) 70 percent commonality for all of its eight vehicles in the 16-ton category.[6] A key parallel was drawn between the FCS and UAS procurement efforts. Even though the eight FCS variants include a variety of capabilities from a non-line of sight cannon to a medical and evacuation system, all utilize the same chassis.[7] Consolidation of medium and high altitude UAS platforms follows the same airframe principle in order to increase interoperability and reduce logistical requirements.

In order to apply the 75 percent threshold to this study, a formula was created to accurately reflect the capability similarities between UAS platforms. This percent value results from comparing the capabilities of two UAS and identifying the capability numbers to be compared such as speed, weight or time. The percentage of similar capability results by dividing the lower UAS capability number by the higher UAS capability number. Any two or more systems that meet the 75 percent similarity threshold would be highlighted. Upon conclusion of the quantitative analysis, any UAS systems that shared the similarity threshold in three or more of the capability areas would be identified for further quantitative analysis.

$$\frac{\text{Lower UAS Value (x)}}{\text{Higher UAS Value (y)}} = \text{Percentage of Similar Capability (z)}$$

Figure 4.   Similarity Formula

The second UAS component essential to this research are the payload considerations of all five UAS. Qualitative analysis on the UAS payload considerations confirm or reject the quantitative results of the statistical capability comparisons. UAS payload allocation can vary significantly based on different service requirements and aerodynamic capability, so seven payload considerations were chosen: internal and external payload, fuel type, weaponization, imagery, signals and radar sensors. This analysis offered no numerical indications, but did confirm or dispel the feasibility of merging UAS programs based on the feasibility of integrating respective payloads.

Upon completion of these research methodologies, analysis should produce findings from which recommendations can be made. By determining the differences in magnitude of funding and fiscal percentage of the five UAS systems, the significance of UAS related investments will be apparent. Through the quantitative capability analysis and qualitative payload comparisons, evidence will reveal which UAS programs exhibit duplicative capabilities and show possibilities of platform consolidation. Operational risk analysis will incorporate all the findings and determine what, if any, impacts can be expected from consolidation of the most similar systems. The Air Force definition of operational risk is the impact to mission effectiveness at all levels.[8] This thesis addresses operational risk as any degradation to UAS effectiveness and logistical processes at the operational and tactical levels. Based on the outcome of the research methodology and analysis, a recommendation can be made on whether a single acquisition authority would be in the best interest of the DOD. Addressing the secondary research questions, recommendations will also be made on which systems show promise on potential platform consolidation.

Strength and Weaknesses of Methodology

There are several strengths to the chosen methodology and focus areas of this study. First, comparing fiscal data for all five UAS across the same FYDP required no conversion of current day value. This reduced the error and potential skewing of comparative financial data among different UAS programs. Similarly, all five UAS are compared during the same points in time across the FYDP. Both funding and mission capabilities were easily compared and contrasted in the same reference of time and operating environment. This reduces the influence of outside systems when this study compared two or more different points in time.

An additional strength of this study is the analysis of programmed funding at the according to the 2009 President's Budget. This ensured an accurate representation of the Defense Department's true focus and intentions. Anything higher would have allowed Legislative Branch funding influences to affect the original DOD goals for UAS programs.

The weaknesses in this methodology are discrepancies in the stated performance capabilities of UAS systems from DOD and contractor data sets. The methodology had to account for potential conflicts in performance statistics. Whether due to different testing environments and conditions or inflated claims, the numbers did not always match exactly. Thresholds of deviation had to be scrutinized to determine if additional analysis was required utilizing both data sets in order to identify the implications of stated differences. In cases where two sets of data required analysis, a caveat had to be associated with the findings to normalize both sets of data.

<u>Summary</u>

It was important to establish a methodology that began with funding significance to determine the UAS systems that would benefit the most from a single acquisition authority. The methodology would then target capability redundancy to confirm or reject the notion of consolidation of similar UAS. With the information gathered and the framework of this UAS study established, efforts of this thesis transition to the analysis of data. Results from this study's analysis will be used to make recommendations on the primary and secondary research questions.

---

[1]General Atomics, —Arcraft Platforms," http://www.gaasi.com/products/index.php (accessed 1 February 2009).

[2]Northrop Grumman, —Unmanned Systems," http://www.is northropgrumman.com/ systems/systems_ums html (accessed 11 February 2009).

[3]Ibid.

[4]Ibid.

[5]Defense Acquisition University, —Operations and Maintenance Funds," https://acc.dau.mil/CommunityBrowser.aspx?id=28980 (accessed 4 March 2009).

[6]National Defense, —Future Combat System Begins to Take Shape," http://www.nationaldefensemagazine.org/ARCHIVE/2002/JULY/Pages/Future_Combat4041.aspx (accessed 2 May 2009).

[7]Armed Forces Communications and Electronics Association, —Future Combat Systems makes the grade," http://www.afcea.org/signal/articles/templates/SIGNAL_Article_Template.asp?articleid= 1662&zoneid=40 (accessed 3 May 2009).

8US Air Force, *Air Force Pamphlet 91-216, USAF Safety Deployment and Contingency Pamphlet* (Washington, DC, Government Printing Office, 2001), 1.

CHAPTER 4

ANALYSIS

The fiscal analysis will show the importance of magnitude and trends associated with the five selected medium and high altitude UAS. A capability analysis will identify any areas of duplication as well as potential areas of consolidation. Next, a qualitative analysis on payload consideration will validate the feasibility of merging highlighted programs. Finally, analysis focused on operational risk associated with prospective consolidation will end this chapter.

Fiscal Analysis

Why create a single acquisition entity for the procurement of select medium and high altitude systems today, when it had already failed in the late 1980s and early 1990s? The difference between 1988 UAV issues and 2009 UAS issues are purely a matter of scale. The amounts of UAS funding involved and the proportion of daily operational involvement are polar opposites. To illustrate, this section will identify the significance of UAS growth and the influence of the five platforms on the entire DOD UAS budget.

With UAS identified as the primary target for this study's analysis, key programs needed to be singled out for further analysis. In FY09, the Defense Department programmed $3,271.8 billion for the research and procurement of all UAS projects. Of that amount, the five UAS in this study constitute $2,271.8 billion of the allocated funds, or 69.4 percent of all UAS investment dollars.[1] Figure 5 below depicted the funding break out in the FYDP.

It is important to note that these five systems share prime contractors; General Atomics produced the MQ-1 Predator and MQ-9 Reaper for the Air Force and the MQ-1C Sky Warrior for the Army.[2] With General Atomics receiving 24 percent of the DOD UAS funding, defense leadership could have focused attention on potential cost savings in one defense contract versus two contracts with the same vendor. An acquisition authority could have worked to merge some of these three vehicles, or at least renegotiate with General Atomics for a single contract. Especially since all three of these UAS are in the medium altitude category, share similar missions, capabilities and contractors.

The same logic holds true for the Global Hawk and BAMS programs. With two identical air vehicles, there should have been one contract to Northrop Grumman that represented the requirements of both the Air Force and Navy. The amount of funding at stake between these two programs constitutes $1.477 billion or 45 percent of the entire DOD UAS budget for FY09 alone.[3] Although the Air Force took the initiative to work in coordination with the Navy during the testing that led up to the BAMS contract award, there was no mandate for the Air Force to assist.[4]

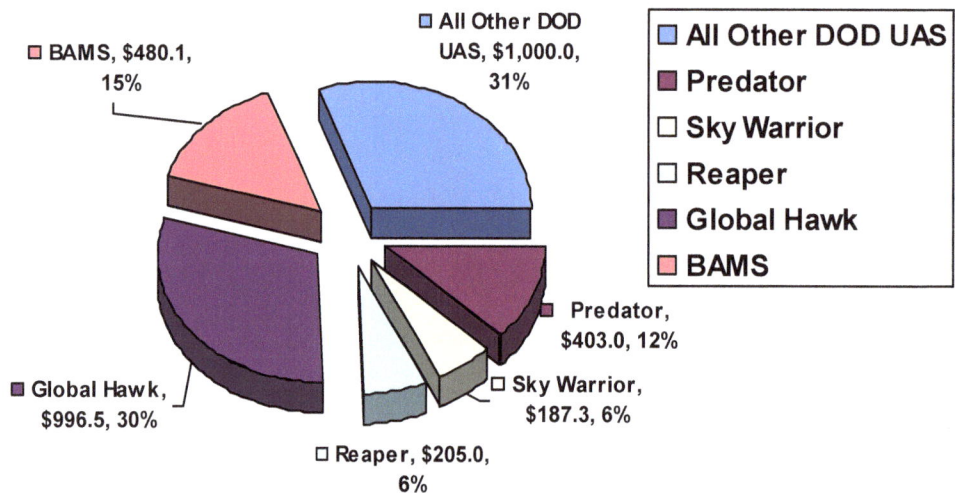

Figure 5.   FY09 Research and Procurement Funding for All DOD UAS Programs
(Dollars in Millions)
Source: Office of the Secratary of Defense, *Unmanned Systems Roadmap 2007-2032*,
Roadmap, Washington DC: Government Printing Office, 2007, 10.

Nevertheless, the actual procurement contracts are still two contracts. An

acquisition authority could have combined these two contracts into one and reaped the

benefits from economies of scale. By spreading fixed costs such as research, factories,

equipment and management over a larger number of Global Hawk airframes, unit costs

would be expected to be lower for both respective services.[5] Utilizing the economy of

scale formula of Average Costs = Total Costs / Quantity, a rough estimate of savings can

be extrapolated.[6] Per the limitations in chapter 2, this equation cannot be performed as

the total costs and total quantity for BAMS are not yet known; initial procurement is not

planned until 2014, which falls outside the current FYDP.[7] However, the average cost

for a Global Hawk is $160.7 million: $9.8 billion total cost divided by 61 aircraft.[8] With

an identical airframe, as the number of procured BAMS aircraft increase, a linear cost savings could be expected. Theoretically, if 30 BAMS were procured at the same average cost of $160.7 million, the total cost for both programs would be approximately $15 billion. If only a 5 percent economy of scale saving was applied, this would equate to $750 million in savings. If an equal number of 61 BAMS were procured, that 5 percent economy of scale saving would equate to $1.5 billion in savings.

In order to analyze potential savings, several areas must be considered. Positive attributes that surrounded the Predator and Warrior programs included the same prime contractor, same site assembly, similar mission type and economy of scale benefits. The RQ-4 Global Hawk and BAMS also share the same prime contractor, similar mission type and potential for economy of scale benefits. This, in addition to the fact that Global Hawk and BAMS share the exact same air vehicle, should increase cost savings above the $400 - $600 million projected for dissimilar air vehicles in the Predator and Warrior programs.[9] The Global Hawk and BAMS programs are 2.5 times the investment amount of the Predator and Sky Warrior programs in FY09.[10] An extrapolation of this ratio led to potential savings in the $1.0 - $1.5 billion range through the consolidation of the Global Hawk and BAMS high altitude UAS programs. This reinforces the theoretical economy of scale example mentioned previously. If properly managed, these four UAS procurement efforts could save over $2 billion dollars over the lifetime of all four programs. Due to the magnitude of funding, similar mission profiles and identical contractors that the MQ-1 Predator, MQ-1C Sky Warrior, MQ-9 Reaper, RQ-4 Global Hawk and BAMS were selected as the five systems for further capability analysis.

Capability Analysis

In light of the coordination and duplication concerns revealed by the GAO, the aircraft capabilities of the five selected UAS were compared with each other. Five key performance areas were chosen which represent the physical capability limits airframes. The categories of maximum altitude, speed, endurance, radius and payload were utilized for the following comparative graphs. The attribute of fuel capacity was rejected as it is not a direct indicator of UAS performance.

The study attempted to identify any natural similarities that occurred repeatedly between any of the 5 UAS systems. A threshold of 75 percent similarity was determined to identify system similarities in capability. If any platforms met the established 75 percent level of commonality in any area, this highlighted a potential opportunity for the DOD to consider merging of multiple programs to eliminate capability duplication and increase cost savings. The more areas that shared this 75 percent level of commonality reflected more rationale to combine the respective UAS.

Table 2.   Medium and High Altitude UAS Capability Comparison

	MQ-1B Predator	MQ-1C Sky Warrior	MQ-9 Reaper	RQ-4B Global Hawk Block 30/40	BAMS*
Initial Operating Capability (IOC)	1995	2012	2007	2001	2015
Maximum Altitude	25,000 ft	29,000 ft	50,000+ ft	60,000 ft	60,000 ft
Maximum Air Speed	138 kts	155 kts	276 kts	340 kts	340 kts
Maximum Endurance	24 hrs	40 hrs	24 hrs	28 hrs	28 hrs
Radius	500 nm	648 nm	1,655 nm	5,400 nm	5,400 nm
Internal Payload Capacity	450 lbs	575 lbs	750 lbs	3,000 lbs	3,000 lbs
External Payload Capacity	300 lbs	500 lbs	3,000 lbs	N/A	N/A
Fuel Capacity	640 lbs	600 lbs	4,000 lbs	16,320 lbs	16,320 lbs

Source: Office of the Secratary of Defense, *Unmanned Systems Roadmap 2007-2032* (Washington, DC: Government Printing Office, 2007), 65-66, 68-69, 73; General Atomics, —Arcraft Platforms," http://www.gaasi.com/products/index.php (accessed 1 February 2009).

Maximum altitude indicates the general height a UAS would be required to attain in order to conduct its mission. Medium altitude UAS tend to operate at altitudes that provide a tactical advantage and still allow weapon engagements in a short amount of time. High altitude UAS usually operate at loftier heights to allow sensors to cover the widest area possible as they conduct operational and strategic level missions. In the comparison of maximum altitudes, three groupings emerged within the 75 percent similarity window. The MQ-1 Predator and MQ-1C Sky Warrior shared an 86.2 percent capability level. The RQ-4 Global Hawk and BAMS shared a 100.0 percent capability level, which is no surprise as they currently utilize the same air vehicle. Finally, the MQ-

9 Reaper shared an 83.3 percent commonality with the Global Hawk and BAMS with respect to altitude.

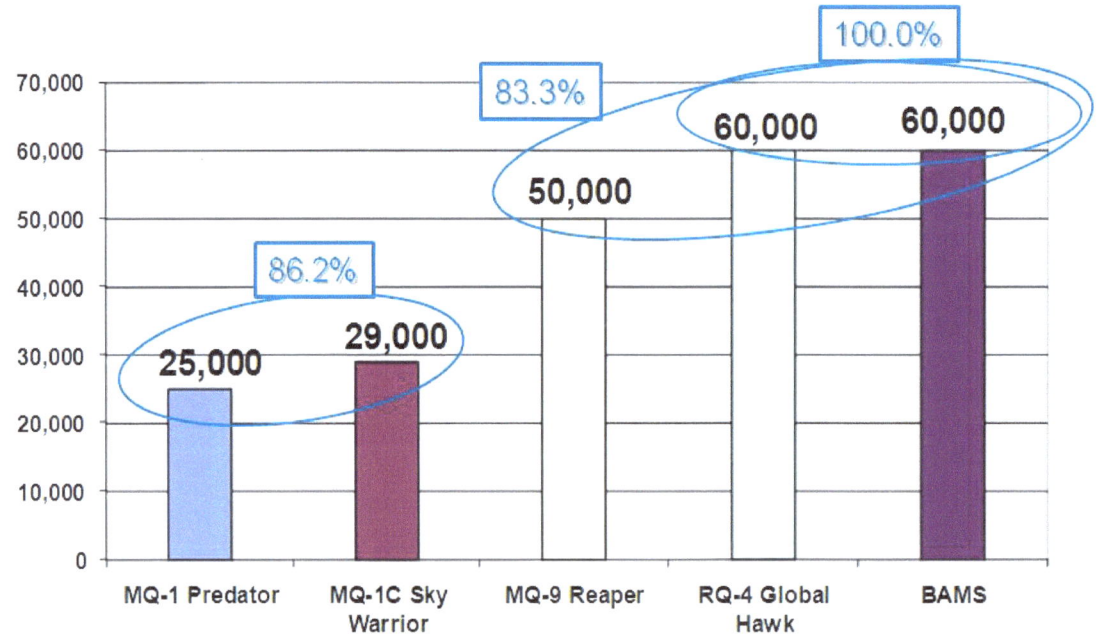

Figure 6.   Maximum Altitude Comparison (Altitude in Feet)
Source: Office of the Secratary of Defense, *Unmanned Systems Roadmap 2007-2032* (Washington DC: Government Printing Office, 2007), 65-66, 68-69, 73.

Maximum speed indicates the comparative response times of the respective UAS. Whether reacting to troops in contact, a time sensitive target or an immediate intelligence re-tasking, speed provides the warfighter more capability to quickly transit the area of operation. When maximum speeds were compared, three UAS groupings showed close similarity. First, the MQ-1 Predator and MQ-1C Sky Warrior shared an 89.0 percent capability level. The RQ-4 Global Hawk and BAMS again resulted in a 100.0 percent similar capability level. Lastly, the MQ-9 Reaper shared an 81.2 percent capability similarity with both the Global Hawk and BAMS.

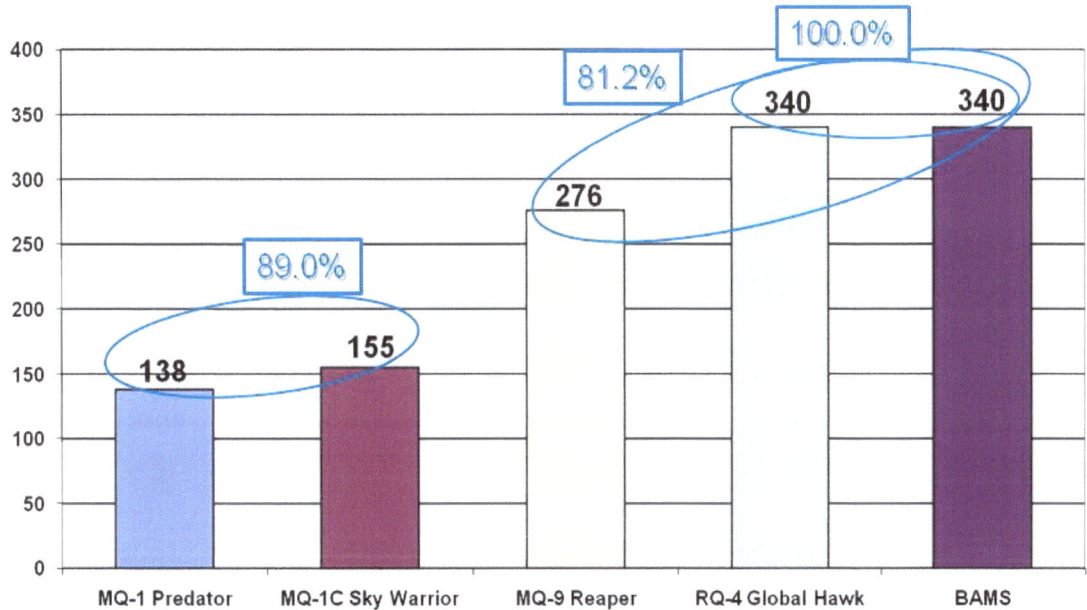

Figure 7.   Maximum Speed Comparison (Speed in Knots)
Source: Office of the Secratary of Defense, *Unmanned Systems Roadmap 2007-2032* (Washington, DC: Government Printing Office, 2007), 65-66, 68-69, 73.

Maximum endurance dictates the length of time a UAS could stay airborne before returning or being replaced by another UAS to continue persistent coverage. UAS with longer endurance required fewer unmanned aircraft to sustain a period of 24 hour coverage. In this study of maximum endurance, the results were not as clear. According to DOD statistics, the MQ-1 Predator and MQ-9 Reaper share a 100 percent capability level as did the RQ-4 Global Hawk and BAMS. In addition, the Predator and Reaper share an 85.7 percent capability level with the Global Hawk and BAMS. The MQ-1C Sky Warrior is the highest outlier at 40 hours and did not meet the 75.0 percent similarity threshold with any system. It shared a 70 percent capability level with the Global Hawk

and BAMS and only 60 percent level of similarity compared to the Predator and Reaper.

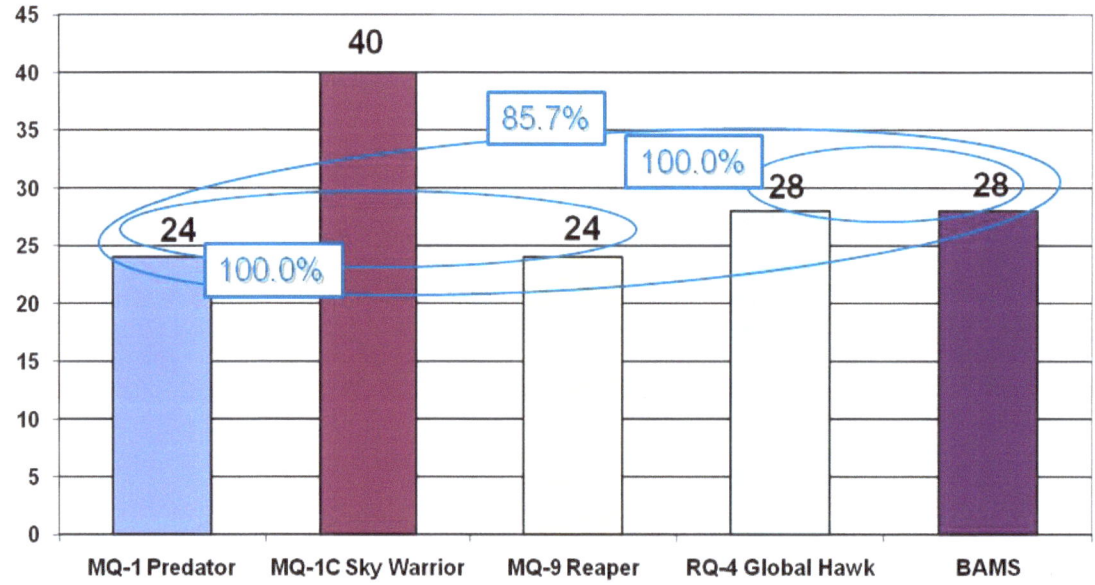

Figure 8.   Maximum Endurance Comparison (Time in Hours)
Source: Office of the Secretary of Defense, *Unmanned Systems Roadmap 2007-2032*
(Washington, DC: Government Printing Office, 2007), 65-66, 68-69, 73.

In contrast to this comparison of aerial endurance, General Atomics, the producer

of the MQ-1C Sky Warrior, claimed it only had an endurance of 32 hours vice 40 hours.

This was a performance difference of 20 percent from the DOD listed capability.

Utilizing the contractor data set, the Sky Warrior now shared an 87.5 percent capability

level with the Global Hawk and BAMS, and a 75.0 percent capability level with the

Predator and Reaper, indicating that the Sky Warrior met the 75.0 percent similarity

threshold with all the other UAS. It is important to note the vendor provided

performance data produced the exact opposite quantitative results compared to DOD performance data. The impact is minimal in this situation as the contractor did not

exaggerate performance claims or attempt to offer a vastly different UAS from an

endurance perspective. The contractor viewed the Sky Warrior's endurance as an

incremental improvement over the Predator and Reaper.

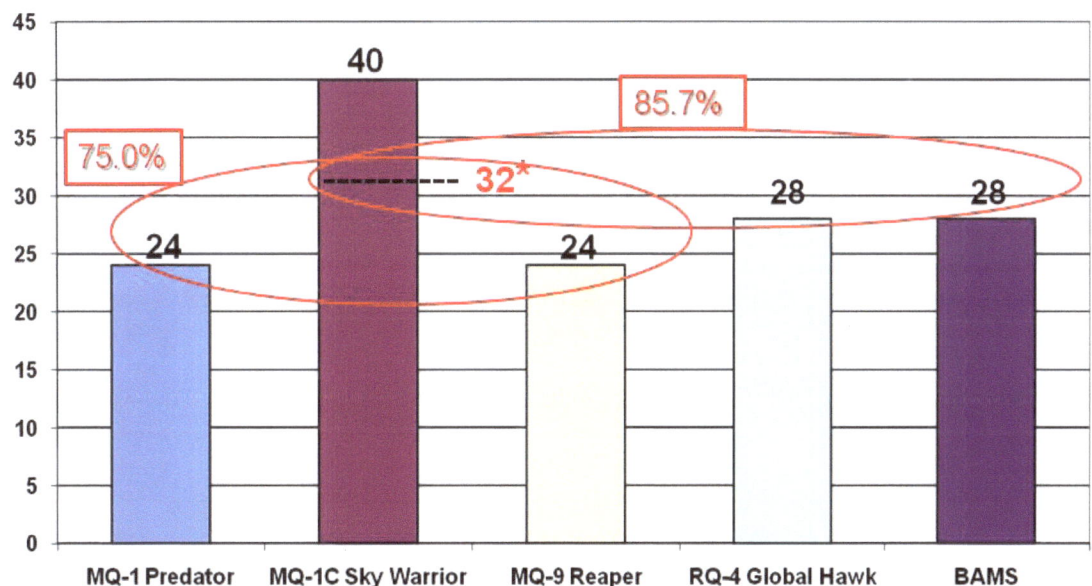

Figure 9.   Maximum Endurance Comparison with Contractor stated Endurance (Time in
Hours)
Source: Office of the Secratary of Defense, *Unmanned Systems Roadmap 2007-2032*,
(Washington, DC: Government Printing Office, 2007), 65-66, 68-69, 73; General
Atomics, ―Arcraft Platforms," http://www.gaasi.com/products/index.php (accessed 1
February 2009). *Note*: *General Atomics listed maximum endurance of the MQ-1C Sky
Warrior at 32 hours

Maximum radius indicates the farthest distance an aircraft can travel and return to

its point of departure. A larger maximum radius indicates more area an aircraft can cover

which could influence military operations. Comparisons of maximum radius between

systems identified two similar groupings. The Predator and Sky Warrior were grouped

with a 77.2 percent similarity. Global Hawk and BAMS received a 100.0 percent commonality as in previous graphs. The Reaper didn't fit into either group with a 30.6

percent commonality with the high altitude UAS and a 39.1 percent commonality with

the Sky Warrior.

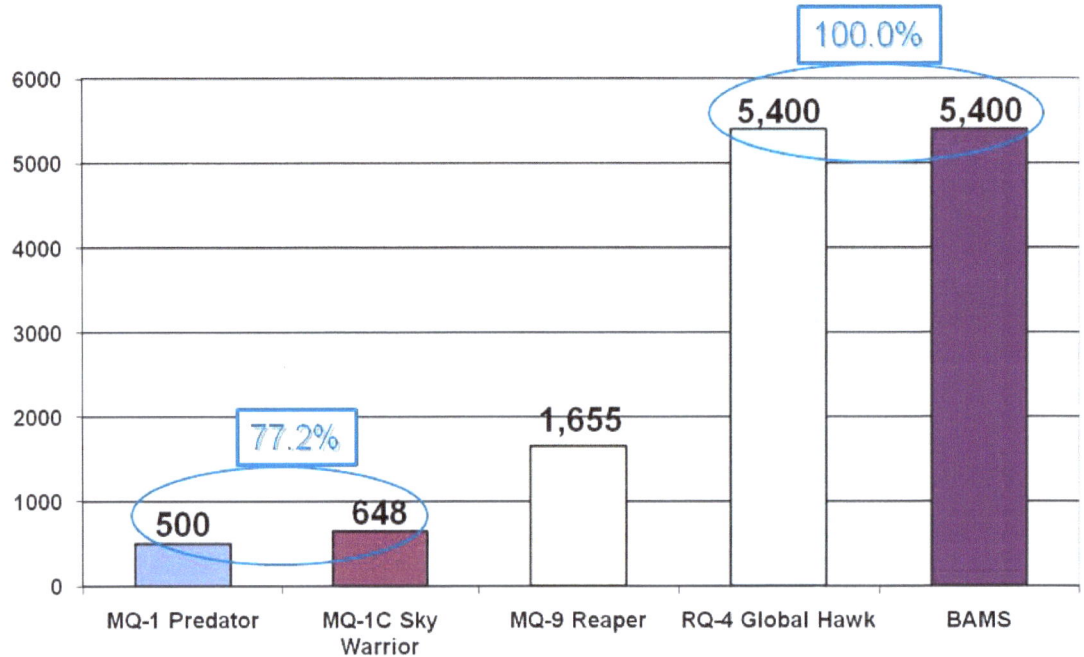

Figure 10. Maximum Radius Comparison (Distance in Nautical Miles)
Source: Office of the Secratary of Defense, *Unmanned Systems Roadmap 2007-2032*
(Washington, DC: Government Printing Office, 2007), 65-66, 68-69, 73.

The comparison of maximum payload highlights the lift capability of each

respective UAS. A high maximum payload indicates the amount of sensors,

communication and weapons payload carried by the respective system. This final

category revealed two distinct groupings. The medium altitude MQ-9 Reaper shared an

80.0 percent payload commonality with the high altitude RQ-4 Global Hawk and BAMS.

The Global Hawk and BAMS displayed a 100.0 percent similarity in capability. The

MQ-1 Predator and MQ-1C Sky Warrior did not meet the 75 percent threshold with only

a 69.8 percent rating between the two systems.

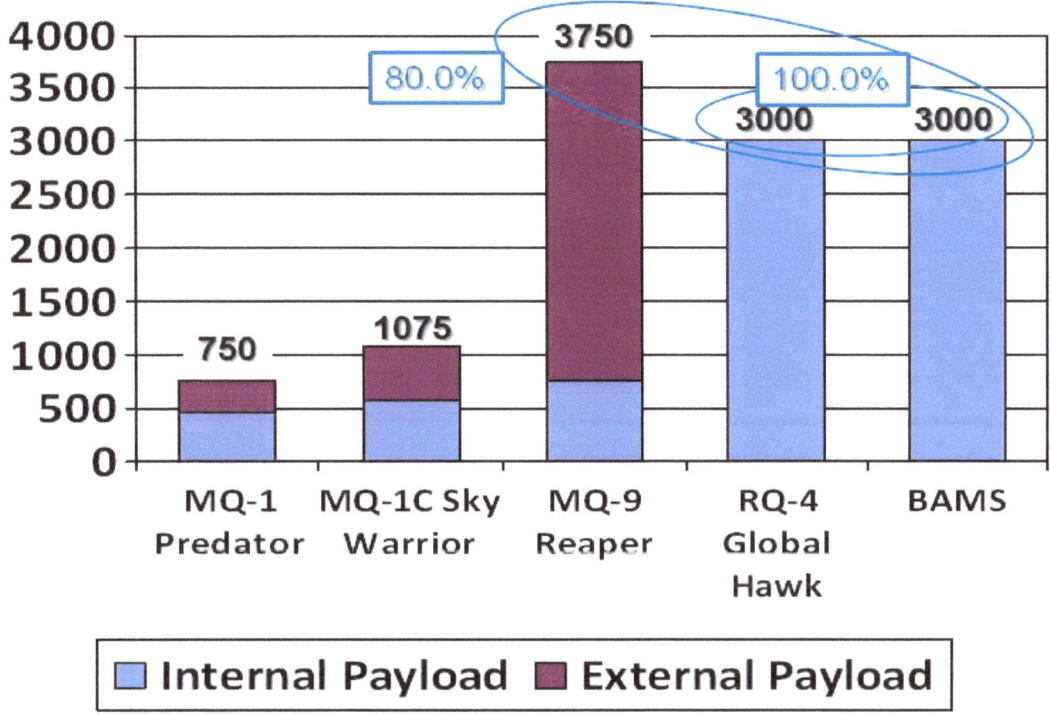

Figure 11. Maximum Payload Comparison (Internal and External) (Weight in Pounds)
Source: Office of the Secratary of Defense, *Unmanned Systems Roadmap 2007-2032*
(Washington, DC: Government Printing Office, 2007), 65-66, 68-69, 73.

Three distinct groups emerged from the capability analysis that showed very

similar tendencies. First, the MQ-1 Predator and MQ-1C Sky Warrior met the 75.0

percent similarity threshold in three of the five capability areas. They met the threshold

in four of the five capability areas when contractor data was utilized instead of DOD

endurance data. Only maximum payload did not meet the similarity threshold in with

either data set. This grouping will be identified as Group 1.

The second distinct grouping highlighted that the MQ-9 Reaper shared four of the five similarity thresholds with the RQ-4 Global Hawk and BAMS. Maximum radius was the only category that did not meet the similarity threshold. This grouping will be identified as Group 2.

The final grouping indicated that the RQ-4 Global Hawk and BAMS was a perfect match. They not only met the 75 percent similarity threshold in all five capability areas, but met this with 100.0 percent commonality in every trait. This grouping will be identified as Group 3.

With three separate groupings identified out of all five UAS compared, a different analytical process was applied to the three groupings. The next section takes a narrower and qualitative look at the distinct payload considerations of the three groupings identified.

Payload Considerations

The details of the payload differences and similarities were analyzed to provide a qualitative perspective on potential commonality between the three UAS groupings identified in the previous section. In the table below, four distinct clusters displayed similar traits within the three groups under consideration.

In Group 1, the MQ-1 Predator and MQ-1C Sky Warrior did not meet the 75 percent similarity threshold for payload in the quantitative analysis, but came close with a 69.8 percent commonality; but upon further analysis, both have a similar ratio of payloads. The Predator held 60.0 percent of its payload internally and 40 percent externally. The Sky Warrior held 53.5 percent of its payload internally and 46.5 percent externally. This similarity in payload distribution shows potential for consolidation as both require a balance between internal sensors and external weapons.

The MQ-1 Predator and MQ-1C Sky Warrior also shared commonality in the types of payload carried. Externally, both utilized the AGM-114 Hellfire as its only weapon system. Internally, each system carried very similar suites of imagery, signal and radar sensors.

In Group 2, the MQ-9 Reaper, RQ-4 Global Hawk and BAMS displayed identical requirements in the type of fuel required. JP-8 was the propulsion fuel for these three systems. But that was the only qualitative similarity shared between the Reaper and the high altitude Global Hawk and BAMS.

In Group 3, the RQ-4 Global Hawk and BAMS meet identical qualitative comparisons of payload. Internal payload amount, proportion of internal and external payload and fuel type were exact matches. Neither system possessed a weapons delivery capability. Imagery and signals sensors, although unknown for the BAMS, would be comparable with the Global Hawk sensors and not require significant changes to the aerial vehicle.

Table 3.   Areas of Potential Payload Consolidation

	MQ-1B Predator	MQ-1C Sky Warrior	MQ-9 Reaper	RQ-4B Global Hawk Block 30/40	BAMS*
Internal Payload	450 lbs	575 lbs	750 lbs	3,000 lbs	3,000 lbs
External Payload	300 lbs	500 lbs	3,000 lbs	N/A	N/A
Fuel Type	AVGAS	JP-8	JP-8	JP-8	JP-8
Weaponization	2x AGM-114 Hellfire	4x AGM-114 Hellfire	8x AGM-114 Hellfire 4x GBU-12 LGB 4x GBU-38 JDAM	N/A	N/A
Imagery Sensor	MTS-A EO/IR camera system Full Motion Video	MTS-A EO/IR camera system Full Motion Video	MTS-B EO/IR camera system Full Motion Video	Electro-Optical/Infrared Hi Resolution Stills	TBD
Signals Sensor	SIGINT/ESM	SIGINT	SIGINT/ESM	Yes	TBD
Radar Sensor	Lynx SAR	Lynx SAR/MTI	Lynx SAR/GMTI	Synthetic Aperture Radar (SAR)	TBD (360 degree field of view)

Source: Office of the Secratary of Defense, *Unmanned Systems Roadmap 2007-2032* (Washington, DC: Government Printing Office, 2007), 65-66, 68-69, 73; General Atomics, —Arcraft Platforms," http://www.gaasi.com/products/index.php (accessed 1 February 2009). *Note*: * BAMS performance statistics are notional based off of the Northrop Grumman contract award vehicle (Block 30/40 Global Hawk).

Utilizing the same payload data, a qualitative look at dissimilar payloads identified any problems that might detract from combining UAS programs. Areas that required significant physical change to merge airframes were not conducive to consolidation. In other words, trends in this section negatively impact the potential merging of two or more UAS contracts.

In Group 1, the MQ-1 Predator and MQ-1C Sky Warrior displayed only one area of non-consolidation. The fuel type was not standard as the Predator utilized aviation gas

(AVGAS) while the Sky Warrior used jet propulsion 8 (JP-8). Fuel type not only impacts affects these two platforms, but has second and third order effects that must be considered as well. From a force structure perspective, utilizing multiple fuels could cause additional infrastructure changes to accommodate this requirement. Logistically, multiple fuel types cause additional strain on the sustainment system, versus a force with vehicles that utilize one common battlefield fuel. The fuel difference between the Predator and Sky Warrior created a serious consolidation issue as there is no way to merge either platform to meet this requirement. Any consolidation decision would have to choose one of these UAS vehicles as the future of the DOD force.

In Group 2, there were four separate areas not favorable to consolidation indicated for the MQ-9 Reaper and the RQ-4 Global Hawk and BAMS. Although the quantitative analysis showed an 80.0 percent commonality in maximum payload, a qualitative look displayed the opposite. The total payload of these aircraft was close, but the allocation proved non-compatible. First, the Reaper only possessed 750lbs of internal payload, whereas the Global Hawk and BAMS required 3,000lbs. Secondly, the external payload of the Reaper required 3,000lbs, while the Global Hawk and BAMS possessed no capability to incorporate this requirement. The third area on non-consolidation was the weapons ability of these systems. The Reaper required the use of several missiles and precision guided munitions in order to accomplish its mission. Conversely, the Global Hawk and BAMS possessed no strike capability. The fourth and final area of non-consolidation was the imagery sensors on these systems. The Reaper required real-time, full-motion video to execute its mission. The Global Hawk and BAMS required an imagery sensor capable of capturing still imagery of large areas. The non-consolidation

59

differences between the Reaper and the Global Hawk and BAMS were centered on the disposition of payload. The Reaper was weighted heavy on external payload to accommodate weapons delivery. The entire Global Hawk and BAMS payload were internally located for the protection of sensitive collection systems and the maximization of aerodynamics for high altitude flight.[11]

Although the quantitative analysis strongly favored consolidation of the Reaper, Global Hawk and BAMS, the qualitative analysis strongly discounted the Reaper from consideration. There are serious complications when attempting to combine a weapon focused UAS in the Reaper with a strategic ISR focused UAS like the Global Hawk and BAMS. There is neither enough payload commonality nor the correct internal-external payload distribution to merge the Reaper into the Global Hawk and BAMS program and still accomplish both missions.

In Group 3, non-consolidation was not an issue as the RQ-4 Global Hawk and BAMS utilized identical airframes.[12] The only area that may cause some compatibility issues resides in the signal or radar sensor. The U.S. Navy specified that they required an electronic sensor with a 360 degree capability.[13] It remains to be seen if the technology exists to incorporate a 360 degree capability internal to the BAMS or if significant modifications would be required for the airframe. These two airframes showed the most promise for consolidation in both the quantitative and qualitative analysis conducted in this thesis.

Table 4.   Areas of Potential Payload Non-Consolidation

	MQ-1B Predator	MQ-1C Sky Warrior	MQ-9 Reaper	RQ-4B Global Hawk Block 30/40	BAMS*
Internal Payload	450 lbs	575 lbs	750 lbs	3,000 lbs	3,000 lbs
External Payload	300 lbs	500 lbs	3,000 lbs	N/A	N/A
Fuel Type	AVGAS	JP-8	JP-8	JP-8	JP-8
Weaponization	2x AGM-114 Hellfire	4x AGM-114 Hellfire	8x AGM-114 Hellfire 4x GBU-12 LGB 4x GBU-38 JDAM	N/A	N/A
Imagery Sensor	MTS-A EO/IR camera system Full Motion Video	MTS-A EO/IR camera system Full Motion Video	MTS-B EO/IR camera system Full Motion Video	Electro-Optical/Infrared Hi Resolution Stills	TBD
Signals Sensor	SIGINT/ESM	TBD	SIGINT/ESM	Yes	TBD
Radar Sensor	Lynx SAR	Lynx SAR/MTI	Lynx SAR/GMTI	Synthetic Aperture Radar (SAR)	TBD (360 degree field of view)

Source: Office of the Secretary of Defense, *Unmanned Systems Roadmap 2007-2032* (Washington, DC: Government Printing Office, 2007), 65-66, 68, 73; General Atomics, ―Aircraft Platforms,‖ http://www.gaasi.com/products/index.php (accessed 1 February 2009). *Note*: * BAMS performance statistics are notional based off of the Northrop Grumman contract award vehicle (Block 30/40 Global Hawk).

Operational Risk

The findings resulting from this analysis do carry some operational risk. This study's analytical effort concludes with a final qualitative analysis on the risks to users resulting from consolidation decisions within the three similarity groups. Consolidation risks are viewed from both merge scenarios within each identified grouping.

In group 1, risks associated with merging the MQ-1 Predator into the MQ-1C Sky Warrior primarily affect the Air Force. Being the newer system, the Sky Warrior is the next logical evolution to the 15 year old Predator. There would be no risk associated for the Army in this scenario. If anything, the Army would actually pay less for each Sky Warrior based on economies of scale from a joint procurement strategy with the Air Force. The Air Force on the other hand would gain a more capable system with several short-term risks associated with this decision. First, current Predator pilots, sensor operators and maintenance personnel would all require new training to implement the Sky Warrior. Proficiency levels would decrease initially, but increase over time. This will have a negative impact on current operations in OEF and OIF until personnel experience with the Sky Warrior matches current proficiency standards. Secondly, the Air Force will have to remove the AVGAS logistical infrastructure as they integrate JP-8 as the fuel for the new Sky Warrior. This should not be a large problem for Air Force bases flying both Predator and fixed wing aircraft that already utilize JP-8, but will pose a substantial problem for remote locations where the Predator is the only military asset. These situations will require a significant investment of resources and time to integrate JP-8, potentially affecting vital operations in remote locations until JP-8 integration is fully complete.

When analyzing the potential merge of the MQ-1C Sky Warrior into the MQ-1 Predator airframe, serious risk is associated with this proposition. First, the Sky Warrior outperformed the Predator in all five quantitative capability areas. A consolidation in this direction would significantly decrease the combat capability of all Army units currently employing the Sky Warrior. From radius to speed to armament, the Army would sacrifice significant ISR and offensive capability by procuring the older, smaller and slower Predator. Secondly, the Army would have to introduce AVGAS into their logistical structure which runs counter to their single fuel on the battlefield concept. A secondary, parallel fuel distribution system would have to be established creating more duplication and delays. Ramifications would not only be realized in the delays getting AVGAS to the new Predator assets, but also in the delays to other Army customers whose fuel requirement was superseded by the AVGAS priority. Overall Army combat effectiveness would decrease initially until the fuel process matures. Overall Army financial investments will increase indefinitely as a result of duplicate fuel support systems.

Operational risk is the major reason the Group 2 findings trended against the consolidation of the MQ-9 Reaper and the RQ-4 Global Hawk and BAMS airframes from both perspectives. The impact to the warfighter and decision maker is too severe to compromise the capabilities of these UAS by choosing one common airframe. The risk involved with a Reaper to a Global Hawk merge will be addressed first. Although it is physically possible to internalize weapons in the Global Hawk and BAMS airframes, there would be a sacrifice in payload capacity. External payload possessed by the Reaper is not bound by space, but by airframe lift potential. It would be impossible to take the Reaper's 3,000 pounds of external weapons and 750 pounds of internal equipment and place them both in the 3,000 pound cavity of the Global Hawk and BAMS. Quantitatively, this leaves 750 pounds not accounted for in the merge. Qualitatively the weapons payload is much smaller as modifications to the internal storage area will take up additional space. The introduction of weapons release and bomb bay equipment will take up internal storage area thereby reducing the Global Hawk and BAMS 3000 pound internal capacity. Also, the inability to reshape weapons to maximize internal space

63

further reduces the amount of internal payload usable for weapons required by the Reaper. Although, it is possible to weaponize the Global Hawk and BAMS, this merge would not come close to meeting the Reaper's requirement to employ 3000 pounds of munitions with 750 pounds of vital internal equipment.

Different operational risks are associated with the merge of the RQ-4 Global Hawk and BAMS into the MQ-9 Reaper airframe. From a payload perspective, the Reaper can physically lift the 3,000 pounds of equipment associated with the Global Hawk and BAMS. But the operational risk lies in the placement of sensors and the distance the payload can be carried. The Global Hawk and BAMS carries 3,000 pounds of internally based ISR sensors and communications equipment in order to protect these sensitive systems from atmospheric and environmental damage, to maximize the use of space and to increase the aerodynamics of a high altitude, long endurance airframe. Placing the ISR and communication systems externally on a Reaper would cause three problems affecting the intended effectiveness of the Global Hawk and BAMS mission. First, these sensitive sensors would have to be encased in pods and placed externally under the wings of the Reaper, exposing them to more turbulence. This would degrade the imaging and MTI capabilities due to increased instability of the sensors and receivers located under the Reaper's wings versus the internally and centerline balanced Global Hawk configuration. Secondly, complications associated with separating the current 3,000 pounds of internal equipment into separate pods defeats the purpose of consolidating ISR and communication equipment to leverage common requirements and components to maximize the finite space and weight requirements of an airframe. Finally and most importantly, the maximum radius falls significantly short of the strategic ISR requirements demanded of the Global Hawk and BAMS airframes. Even if the Reaper were able to carry all of the ISR and communication equipment of the Global Hawk and BAMS, their maximum radius of 5,400 miles is still 325 percent longer than the Reaper's. Although the Reaper's 1,655 mile maximum radius is sufficient for most theater requirements, it cannot match the 5,400 mile radial requirement of the Global Hawk airframe that fulfills multiple global requirements.[14]

Operational risk is not a large factor for Group 3 for two reasons. First, the Air Force is currently utilizing the Global Hawk airframe proven in combat since 2001.[15] The Navy will utilize the exact same ISR airframe for their maritime surveillance mission. Second, the Navy is still five years away from fielding its first mission capable BAMS aircraft, so there is no operational impact in the near future from a warfighter perspective.[16] Although, if there was an acquisition authority that combined these two procurement contracts, operational risk for the Air Force would actually decrease, as the per unit cost of the Global Hawk airframe would become lower based on economies of scale. In the unfortunate event of a lost aircraft, it would cost the Air Force less to replace. The level of savings is dependent on the total investment amount and total number of BAMS to be procured; which is still unknown at this time.

The qualitative analysis on operational risk led to findings that were applied against the quantitative capability and qualitative payload analysis. In Group 1, operational risk analysis enhanced the argument that the Sky Warrior presented the least risk of the two consolidation options. It reinforced the results of the quantitative capability and qualitative payload analysis. In Group 2, the operational risk analysis countered the quantitative capability analysis. Even though the Reaper and the Global

Hawk-BAMS pairing meet the 75 percent threshold in all capability categories except maximum radius, this turned out to be a critical requirement. The qualitative payload analysis for Group 2 identified payload configuration as a major detractor for consolidation, but it did not recognize the criticality of maximum radius. In Group 3, operational risk enhanced the rationale to merge the Global Hawk and BAMS systems into a single contract as there are only positive operational impacts at this time.

Summary

The quantitative analysis of fiscal data identified several trends with regards to the scope of medium and high altitude UAS funding. First, medium and high altitude UAS funding rose exponentially from 1988 to the present. Medium and high altitude UAS not only made up a vast majority of the UAS budget, but also accounted for the majority of all unmanned system programs in the DOD.

The fiscal analysis was followed by a quantitative analysis of system capabilities. Of the 5 systems in this study, three distinct groupings emerged as potential candidates for platform consolidation. With the three groupings identified, the final analytical study focused on the qualitative impact of payload considerations. There were some issues for consolidation when payload considerations were analyzed. Finally, operational risk was addressed for each of the groupings. Impacts to effectiveness and responsiveness were analyzed from a qualitative perspective to determine any operational ramifications resultant from airframe consolidation.

Chapter 5 will offer conclusions and recommendations to the primary and secondary research questions. Reinforcing background and analysis compiled throughout this study will justify the recommended actions of this study. This thesis will conclude with recommendations for future research efforts that could be conducted related to the

effective and efficient management of UAS.

---

[1]Ibid.

[2]General Atomics, ─Aircraft Platforms," http://www.gaasi.com/products/index.php (accessed 1 February 2009).

[3]Office of the Secratary of Defense, *Unmanned Systems Roadmap 2007-2032* (Washington, DC: Government Printing Office, 200), 10.

[4]Defense News. "Lt Gen David Deptula: US Air Force Deputy Chief of Staff for ISR," http://www.defensenews.com/story.php?i=3896251 (accessed 12 January 2009).

[5]Linux Project, Economies of Scale Definition, http://www.linfo.org/economies_of_scale.html (accessed 1 May 2009).

[6]Saskatchewan Schools, Economy of Scale, http://www.saskschools.ca/curr_content/socst10_05/unit2/handouts/economies_of_scale.ppt (accessed 10 May 2009).

[7]Office of the Secratary of Defense, *Unmanned Systems Roadmap 2007-2032* (Washington, DC: Government Printing Office, 2007), 77.

[8]Deagal, RQ-4 Global Hawk, http://www.deagel.com/AEWandC-ISR-and-EW-Aircraft/RQ-4A-Global-Hawk_a000556001.aspx; Internet: accessed 10 May 2009.

[9]US Air Force briefing 6 February 2008, *USAF Unmanned Aircraft Systems Vision* (Washington, DC: AF/A2, 2008), 4.

[10]Office of the Secretary of Defense, *Unmanned Systems Roadmap 2007-2032* (Washington, DC: Government Printing Office, 2007), 65-66, 68, 73.

[11]Office of the Secretary of Defense, *Unmanned Systems Roadmap 2007-2032* (Washington, DC: Government Printing Office, 2007), 65-66, 68-69, 73.

[12]Northrop Grumman, ─Unmanned Systems," http://www.is.northropgrumman.com/systems/systems_ums.html (accessed 11 Feb 2009).

[13]Office of the Secretary of Defense, *Unmanned Systems Roadmap 2007-2032* (Washington, DC: Government Printing Office, 2007), 69.

[14]Office of the Secretary of Defense, *Unmanned Systems Roadmap 2007-2032* (Washington, DC: Government Printing Office, 2007), 65-66, 68-69, 73.

[15]Ibid.

[16]Ibid.

CHAPTER 5

CONCLUSIONS AND RECOMMENDATIONS

This thesis did find that economic efficiencies could be achieved by appointing a lead service or agency to direct the procurement of the five medium to high altitude UAS systems under review. It is apparent that UAS are the focus area for DOD unmanned efforts in the present and near-term. Accounting for 89 percent of the DOD budget unmanned, UAS programs have continued to grow exponentially. Significant opportunities exist for DOD to garner cost savings based on the large magnitude of fiscal investments and existing duplication of medium and high altitude UAS procurement and capabilities. Of all the UAS programs in DOD, five UAS accounted for $1.88 billion or 69.4 percent of the entire DOD UAS portfolio in FY09 spread between two defense contractors. Economies of scale analysis revealed that savings resultant from a Predator-Sky Warrior and Global Hawk-BAMS consolidation would be on the order of over $2 billion dollars. A single acquisition authority could have forced a consolidation of UAS programs to gain significant savings despite divergent procurement efforts from the Army, Navy and Air Force.

With the large magnitude of fiscal scale established through funding analysis, areas of inefficiency identified in acquisition and duplication needed to be proven. This was reinforced by the numerous GAO reports highlighting duplication and management concerns in the acquisition of medium and high altitude UAS. This study proceeded to address the specific requirements of the five largest UAS programs in the DOD to see if service UAS contained redundant capabilities. If so, an overall acquisition authority could intervene to merge separate service programs that provide similar capabilities to the warfighter.

In response to the secondary research questions, this thesis also found that there are similar mission requirements that support the potential to combine some UAS into a single system, administered by one agency, service or organization. It also identified some differences between the UAS capabilities justifying divergent acquisition programs. Quantitative comparisons of capability: maximum altitude, speed, endurance, radius and payload, highlighted three groupings that met the 75 percent similarity threshold in three or more performance areas. The MQ-1 Predator and MQ-1C Sky Warrior meet the similarity threshold in three capability areas and were identified as Group 1. The MQ-9 Reaper, RQ-4 Global Hawk and BAMS shared four capability areas and were identified as Group 2. The RQ-4 Global Hawk and BAMS meet the threshold in all five capability areas and were identified as Group 3.

From the comparison of capabilities, the three identified groups were subjected to a qualitative appraisal in regards to the payload considerations and the potential impact to system consolidation. The qualitative analysis in this section focused on payload distribution, fuel type, weapon systems, electro-optical sensors, signals sensors and radar sensors. The following rationale explains why two of the Groupings were supported for consolidation while one was rejected.

Group 1, comprised of the MQ-1 Predator and MQ-1C Sky Warrior, is recommended for program consolidation. They exhibited several similarities throughout the analysis, but did have slightly different maximum payload levels and totally different fuel types in AVGAS and JP-8 respectively. The recommendation of this study would be to merge the MQ-1 Predator program into the MQ-1C Sky Warrior production line. The

Sky Warrior provided more weapons and payload options, increasing flexibility over the

Predator. The Sky Warrior's use of JP-8 made it a logistical benefit with the Army's

–single fuel on the battlefield policy."[1] The M1 Abram, M2 Bradley, HMMWV, heavy

trucks, helicopters and most Army boats currently use JP-8 for fuel.[2] Additionally, JP-8

has become the primary fuel type utilized by the US military and North Atlantic Treaty

Organization (NATO) nations.[3] This single battlefield fuel type would reduce the

sustainment problems associated from utilizing two separate fuels at military operating

bases.[4] Finally, the Sky Warrior had higher performance statistics in all five of the

quantitative capability levels when compared to the Predator. Operational risk analysis

findings were in concurrence with both the fuel and airframe superiority rationale to

merge the Predator into the Sky Warrior program. Operationally, the Sky Warrior

provides the warfighter a more capable UAS platform while avoiding battlefield impacts

from a dual logistical support process.

Group 2, comprised of the MQ-9 Reaper, RQ-4 Global Hawk, and BAMS, are

rejected for program consolidation. Although they meet the 75.0 percent capability in

four of the five quantitative performance areas, the qualitative analysis did not support

merging programs. Payload distribution was a qualitative factor that had no compromise.

Although they shared an 80.0 percent similarity in total payload, the internal and external

distribution did not allow for consolidation. The 3,000lbs of internal payload required for

the Global Hawk and BAMS' sensitive intelligence sensors cannot be placed in the 750

lbs internal payload capacity of the Reaper. Likewise, the 3,000 lbs of external payload

required by the Reaper for a variety of combat munitions cannot be loaded onto the

71

Global Hawk or BAMS as they have no capability for external payloads. Global Hawk and BAMS also differed from the Reaper in imagery capability. Reaper possessed no ability to take still photos of large areas as could the Global Hawk and BAMS platforms. In contrast, the latter systems possessed no full motion video capability like the Reaper. The recommendation for the Reaper is to remain a separate program due to its strike mission and dissimilar payload consideration when compared to the Global Hawk and BAMS. Operational risk analysis concurs with the rejection recommendations, but for different reasons. From an operational perspective, maximum radius was identified as a critical requirement which is not evident from the quantitative capability and qualitative payload analysis. Even if the payload configurations were identical, operational risk analysis would still non-concur with the consolidation as it is impossible for the Reaper to execute the strategic ISR requirements of the Global Hawk-BAMS pair. Operational risk analysis also identified the equally unfeasible task of placing the Reaper's required armament into the Global Hawk's internal cavity.

Group 3, comprised of the RQ-4 Global Hawk and BAMS, are recommended for program consolidation. They exhibited identical quantitative performance statistics in the five measured capabilities. These two UAS also displayed six qualitative areas of potential consolidation and no areas of non-consolidation. The only unknowns were the actual dimensions and fit of the 360 degree surveillance sensor required by the Navy. The consolidation of these two programs was recommended by the DSB back in 2004. The Air Force and Navy had been in coordination throughout the research and test phase of the Navy's new UAS requirement. Operational risk analysis fully supports the recommendation for Group 3 program consolidation. There currently is no negative operational impact to merging both programs into one contract. It is the recommendation of this study that DOD move to consolidate these two systems prior to IOC of the BAMS projected for 2015.[5]

This study concluded that the DOD would benefit from the appointment of one UAS authority. The magnitude of UAS investment, cost savings over $2 billion, and the analytical conclusions that the Predator-Sky Warrior and Global Hawk-BAMS are recommended for program consolidation reinforces the need for one entity to assert their authority and make these decisions now. This UAS authority entity does not necessarily have to dictate every single UAS procurement decision, but it has to have two key attributes. First, this entity must have the visibility to access and understand all DOD UAS programs, both current and projected. Secondly, it must possess the ability and authority to direct UAS platform consolidation where possible and beneficial to the DOD strategic direction. An overarching UAS organization, without the authority to mandate and implement positive UAS changes through funding and program management across services, could perpetuate the poor development and acquisition practices identified by the GAO and the fiscal and capability analysis in this thesis.

Future Considerations

This thesis focused on the acquisition inefficiencies associated with the research and procurement of UAS vehicles. There are several other areas associated with UAS that could be studied for additional areas of funding improvement. Joint basing, operational employment and Congressional language and funding changes all have benefits and drawbacks worthy of analysis to determine the best way forward for the Defense Department.

The final take-away from this study is that unmanned weapon systems are just becoming part of the military's standard force structure. This thesis tried to raise awareness, offer substantive analysis, and provide reasonable recommendations on future DOD acquisition practices. In the future, UAS will be considered as potential options for such roles as multi-role fighter, long range bomber, transporter and aerial refueling in addition to expanding unmanned ground and maritime systems. [6] Hopefully this thesis educated and informed readers on a possible way to make the acquisition process a more effective and efficient way to manage the exploding unmanned domain in its infancy, in order to reap the economic and operational benefits for the next several decades.

_____

[1]Quartermaster Professional Bulletin, ―A Single Fuel for the Battlefield," http://www.quartermaster.army.mil/OQMG/professional_bulletin/1997/Autumn/singlefu el.html (accessed 2 May 2009).

[2]US Army Center for Health Promotion and Preventative Medicine, *JP-8 – Individual* (Aberdeen, MD, Government Printing Office, 2009), 1.

[3]National Center for Biotechnology Information, ―JP-8," http://www.ncbi.nlm. nih.gov/pubmed/17483120?dopt=AbstractPlus (accessed 23 April 2009).

[4]Ibid.

[5] Office of the Secratary of Defense, *Unmanned Systems Roadmap 2007-2032* (Washington, DC: Government Printing Office, 2007), 68-69.

[6]US Congressional Armed Service Committee Conference Report, *Floyd D. Spencer National Defense Authorization Act*, Washington, DC: US Congress, 2001, Section 220.

	FY2007	**FY2008**	**FY2009**
MQ-1B Predator (USAF)			
RDT&E	$ 77.9M*	$ 33.8M	$ 24.8M
Procurement	$ 428.5M	$ 276.1M	$ 378.2M
TOTAL	$ 506.4M	$ 309.9M	$ 403.0M
MQ-1 C Sky Warrior (USA)			
RDT&E	$ 123.7M	$ 44.8M	$ 12.7M
Procurement	$ 38.6M	$ 122.7M	$ 174.6M
TOTAL	$ 162.3M	$ 167.5M	$ 187.3M
MQ-9 Reaper (USAF)			
RDT&E	N/A*	$ 63.9M	$ 43.6M
Procurement	$ 247.6M	$ 58.1M	$ 161.4M
TOTAL	$ 247.6M	$ 122.0M	$ 205.0M
RQ-4 Global Hawk (USAF)			
RDT&E	$ 224.1M	$ 274.7M	$ 284.3M
Procurement	$ 442.6M	$ 580.9M	$ 712.2M
TOTAL	$ 666.7M	$ 855.6M	996.5M
Broad Area Maritime Surveillance (BAMS) (USN)			
RDT&E	$ 26.2M	$ 121.3M	$ 480.1M
Procurement	N/A	N/A	N/A
TOTAL	$ 26.2M	$ 121.3M	$ 480.1M

Figure 12. Recent Investment Spending per UAS
Source: Office of the Secratary of Defense, *Unmanned Systems Roadmap 2007-2032* (Washington, DC: Government Printing Office, 2007), 65-66, 68, 73; Department of the Navy, *Fiscal Year FY2009 Budget Estimates* (Washington, DC: Government Printing Office, 2008), Line item 199. *Note*: *MQ-1 Predator includes MQ-9 Reaper Funding in FY07

BIBLIOGRAPHY

Books

Glade, David B. *Unmanned Aerial Vehicles: Implications for Military Operations.* Maxwell Air Force Base, AL: Center for Strategy and Technology, Air War College, Air University, 2000.

Green, Michael and Greg Stewart. *M1 Abrams at War.* China: Zenith Imprint, 2005.

Leonard, Robert S. and Jeffrey A. Drezner. *Innovative Development: Global Hark and DarkStar: HAE UAV ACTED Program Description and Comparative Analysis.* Vol. 1. Santa Monica, CA: RAND, 2002.

Government Documents

Congressional Budget Office. CBO's Estimate of a Sustaining Budget for National Defense. http://www.cbo.gov/doc.cfm?index=2398&type=0&sequence=3. (accessed 23 April 2009).

————. *Intelligence, Surveillance, and Reconnaissance (ISR). Report to Congress.* Washington DC: Library of Congress, 2005.

————. *Unmanned Aerial Vehicles. Report to Congress,* Washington DC: Library of Congress, 2003.

————. *Unmanned Vehicles for U.S. Naval Forces. Report to Congress.* Washington DC: Library of Congress, 2006.

Curtin, Neal P. GAO-04-530T, Unmanned Aerial Vehicles: Major Management Issue Facing DOD's Development and Fielding Efforts. Washington DC: Government Printing Office, 2004.

Defense Science Board. ―About the DSB.‖ http://www.acq.osd.mil/dsb/ (accessed 1 March 2009.

————. *Unmanned Aerial Vehicles and Uninhabited Combat Aerial Vehicle,* Washington DC: Department of Defense, 2004.

Defense Acquisition Guidebook. ―Joint Capabilities Integration and Development System.‖ https://akss.dau.mil/dag/DoD5000.asp?view=functional (accessed 23 April 2009).

Defense Acquisition University. ―Operations and Maintenance Funds.‖ https://acc.dau.mil/CommunityBrowser.aspx?id=28980 (accessed 4 March 2009).

Department of Defense. ―Contracts." http://www.defenselink.mil/contracts/
contract.aspx?contractid=3758 (accessed 11 February 2009).

———. DOD 5000.01, *The Defense Acquisition System*. Washington DC: Government
Printing Office, 2007.

Department of the Navy. *Fiscal Year FY2009 Budget Estimates*. Washington DC:
Department of the Navy, 2008.

DOD UAS Task Force. Unmanned Aircraft Systems Task Force Briefing, 29 November
2007. Washington DC: Government Printing Office, 2009.

———. Unmanned Aircraft Systems Task Force Briefing, 9 January 2009, Washington
DC: Government Printing Office, 2009.

Government Accountability Office. *Additional Actions Needed to Improve Management
and Integration of DOD Efforts to Support Warfighter Needs. Report to Congress.*
Washington DC: GAO, 2008.

———. *Improved Strategic and Acquisition Planning Can Help Address Emerging
Challenges. Testimony.* Washington DC: GAO, 2005.

———. *Improved Strategic and Acquisition Planning Can Help Address Operational
challenges. Testimony.* Washington DC: GAO, 2006.

———. *Force Structure: Improved Strategic Planning Can Enhance DOD's UAV efforts.
Report to Congress.* Washington DC: GAO, 2004.

———. *UAVs: Major management issues facing DOD's development and fielding
efforts. Testimony.* Washington DC: GAO, 2004.

———. *Progress towards meeting high altitude aircraft pricing goals. Report to
Congress.* Washington DC: GAO, 1998.

———. *DOD UAV Acquisition Efforts. Testimony.* Washington DC: GAO, 1997.

———. Changes in Global Hawk's Acquisition Strategy are needed to reduce program
risks. http://www.gao.gov/products/GAO-05-6 (accessed 30 April 2009).

National Center for Biotechnology Information. ―F-8." http://www.ncbi.nlm.nih.gov/
pubmed/17483120?dopt=AbstractPlus (accessed 23 April 2009).

Office of the Secretary of Defense. *Fiscal Year 2009 Budget Request Summary
Justification*. Washington DC: Department of Defense, 2008.

———. *Unmanned Systems Roadmap 2007-2032.* Roadmap, Washington DC:
Department of Defense, 2007.

Quartermaster Professional Bulletin. "A Single Fuel for the Battlefield."
http://www.quartermaster.army.mil/OQMG/professional_bulletin/1997/Autumn/s
inglefuel.html (accessed 2 May 2009).

Rodrigues, Louis J. Unmanned Aerial Vehicles: DOD's Acquisition Efforts: Statement of
Louis J. Rodrigues, Director, Defense Acquisitions Issues, National Security and
Internationals Affairs Division, Before the Subcommittee on National Security,
House of Representatives (NSIAD-97-138). Washington, DC, General
Accounting Office, 1997.

US Air Force. Air Force Pamphlet 91-216, *USAF Safety Deployment and Contingency
Pamphlet*. Washington, DC: Government Printing Office, 2001.

———. *USAF Unmanned Aircraft Systems Vision*, Briefing 6 February 2008.
Washington, DC: AF/A2, 2008.

———. White paper 6 December 2007. *Joint UAS Operational Employment*.
Washington DC: AF/A2, 2007.

US Army Center for Health Promotion and Preventative Medicine. *JP-8 – Individual*,
Aberdeen, MD, Government Printing Office, 2009.

US Congressional Armed Service Committee Conference Report. *Floyd D. Spencer
National Defense Authorization Act*. Washington DC: US Congress, 2001.

———. *John Warner National Defense Authorization Act*. Washington DC: US
Congress, 2006.

USAF UAV Battlelab. http://www.afc2isrc.af.mil/uav/history.asp (accessed 8 October
2008)

Young, John, 5 October 2007 Memorandum on *Unmanned Aircraft Systems*. Washington
DC: OSD AT&L, 2007.

Other Sources

AM General. "HMMWV Background" http://www.amgeneral.com/vehicles/
hmmwv/background.php (accessed 1 May 2009).

Armed Forces Communications and Electronics Association. "Future Combat Systems
makes the grade." http://www.afcea.org/signal/articles/templates/
SIGNAL_Article_Template.asp?articleid=1662&zoneid=40 (accessed 3 May
2009).

Barnes, Federick. "The Age of UAVs Has Begun, but Manned Aircraft Stay in the
Budget." "Defense & Foreign Affairs Strategic Policy," 30, no. 2 (2002):14-15.

Brown, David. "BAMS (Broad-Area Maritime Surveillance), Eagle Eyes, and Dragon Eyes." *Sea Power* 46 (April 2003): 66-68.

Cahlink, George. "Rise of the Machines." *Government Executive* 36 (July 2004): 76-78.

Deagel. "Equipment Guides." http://www.deagel.com/platforms (accessed 17 April 2009).

———. "RQ-4 Global Hawk." http://www.deagel.com/AEWandC-ISR-and-EW-Aircraft/RQ-4A-Global-Hawk_a000556001.aspx (accessed 10 May 2009).

Defense News. "Lt Gen David Deptula: US Air Force Deputy Chief of Staff for ISR." http://www.defensenews.com/story.php?i=3896251 (accessed 12 January 2009).

Fox News. "Obama Calling for Defense Budget Cuts." http://www.foxnews.com/politics/2009/01/30 (accessed 21 April 2009).

General Atomics. "Aircraft Platforms." http://www.ga-asi.com/products/index.php (accessed 1 February 2009).

Iannotta, Ben. "Who should fly?" *C4ISR Journal* (August 2008): 14-16.

Linux Project. Economies of Scale Definition. http://www.linfo.org/economies_of_scale.html (accessed 1 May 2009).

National Defense. "Future Combat System Begins to Take Shape." http://www.nationaldefensemagazine.org/ARCHIVE/2002/JULY/Pages/Future_Combat4041.aspx (accessed 2 May 2009).

Northrop Grumman. "Broad Area Maritime Surveillance (BAMS)." http://www.irconnect.com/noc/press/pages/news_releases.html?d=140693 (accessed 11 February 2009).

———. "Unmanned Systems." http://www.is.northropgrumman.com/systems/systems_ums.html (accessed 11 February 2009).

Saskatchewan Schools, Economy of Scale. http://www.saskschools.ca/curr_content/socst10_05/unit2/handouts/economies_of_scale.ppt (accessed 10 May 2009).

www.ingramcontent.com/pod-product-compliance
Lightning Source LLC
Chambersburg PA
CBHW050733180526
45159CB00003B/1218